AF262589

UnPlug

UnPlug

A Radiologist Explores the Damage Caused by Electropollution and How You Can Prevent It

ROB BROWN, MD

Foreword by Devra Davis, PhD, MPH

Skyhorse Publishing

Copyright © 2026 by Rob Brown, MD

All Rights Reserved. No part of this book may be reproduced in any manner without the express written consent of the publisher, except in the case of brief excerpts in critical reviews or articles. All inquiries should be addressed to Skyhorse Publishing, 307 Fifth Avenue, 4th Floor, New York, NY 10016.

Children's Health Defense books may be purchased in bulk at special discounts for sales promotion, corporate gifts, fund-raising, or educational purposes. Special editions can also be created to specifications. For details, contact the Special Sales Department, Skyhorse Publishing, 307 Fifth Avenue, 4th Floor, New York, NY 10016 or info@skyhorsepublishing.com.

Children's Health Defense Books® and Skyhorse Publishing® are registered trademarks of Skyhorse Publishing, Inc.®, a Delaware corporation.

Visit our website at www.skyhorsepublishing.com.

Please follow our publisher Tony Lyons on Instagram @tonylyonsisuncertain.

10 9 8 7 6 5 4 3 2 1

Library of Congress Cataloging-in-Publication Data is available on file.

Cover design by David Ter-Avanesyan
Cover image credit: Getty Images

Print ISBN: 978-1-64821-227-7
Ebook ISBN: 978-1-64821-228-4

Printed in the United States of America

To the memory of my father, Barry David Brown, aka "Bee Squared," an electrical engineer who shared his fascination for technology and passion for health and wellness, inspiring me to think the same.

CONTENTS

FOREWORD

There are books that explain, and books that illuminate. And then there are the rare books–this one among them–that do both with equal grace. Rob Brown, physician, radiologist, scientist, and lifelong student of the invisible forces that animate our world, has crafted a work that bridges the gap between the seen and the unseen, between what we believe we understand and what our biology has not yet fully revealed.

Rob possesses a singular combination of gifts. He has the grounding of a trained radiologist, the analytical mind of a clinician who has spent decades studying the human body from the inside out, and the storytelling instincts of a writer who understands that science is not merely a collection of facts but a lens through which we learn to see. With these tools, he guides readers through one of the most consequential scientific issues of our time: our growing immersion in industrially crafted electromagnetic fields.

From the first pages, Rob welcomes us into a journey that is both deeply personal and broadly universal. He has cared for patients as their radiologist, comforted loved ones as a son, and marveled at life as a father. These roles converge in a simple but profound question: *What happens when the modern electrical world intersects with the delicate electrical architecture of the human body?* It is a question that threads through every chapter of this book.

The Invisible Sea We Swim In

Before electricity powered our cities, before the first telegraph wire hummed, before a single human voice was carried across the air, the Earth was already alive with electromagnetic rhythm. Our planet spins inside a vast cocoon of natural fields; our bodies evolved to the gentle cadence of Schumann resonances, the faint radio murmurs left over from the birth of the universe itself. Life has always been electrical at its core.

But in just a few generations–a blink in evolutionary time–we have filled our homes, workplaces, streets, and even our children's cribs with artificial EMF emissions millions of times stronger and infinitely more complex than anything found in our natural world. We carry radiating devices against our hearts, our reproductive organs, our developing bodies. We live in what Rob aptly describes as an electromagnetic "weather system" of our own making. To imagine that such a transformation might be biologically inconsequential is to misunderstand both biology and evolution. Every organism, every cell, every heartbeat listens to the world through electrical language. And when the language changes, the body cannot help but notice.

Electricity as the Essence of Life

Rob writes with a clarity that is both lyrical and startlingly direct. The difference between life and death, he reminds us, is not chemical–it is electrical. Our hearts beat because of coordinated electrical impulses. Our thoughts arise from waves of charged particles. Our circadian rhythms signaling when to sleep and when to rise depend upon electromagnetic cues. When the brain stops generating electrical activity, life ends.

At death, the body contains every molecule it held an instant earlier. What is gone is the electrical symphony that once organized every part of our body. With this in mind, understanding how external electromagnetic fields interact with the internal electrical fabric of the human body is not merely an academic exercise; it is essential to public health, child development, our capacity to have healthy children if and when we choose to do so, and the very future of medicine itself.

A Discovery in Real Time

Rob's own research marks a radical turning point in this conversation. Using the everyday tools of his profession–ultrasound imaging and a clinician's relentless curiosity–he captured, in real time, a biological response to microwave radiation that had never before been documented in a living human subject.

The experiment was simple: place a cell phone behind the knee of a healthy volunteer for five minutes. Then watch what the blood does.

Through the ultrasound screen, Rob observed the red blood cells beginning to gather into tiny vertical stacks resembling towers of coins. Scientists call this formation *rouleaux*. Physicians recognize it as a sign that blood viscosity has temporarily increased and that microcirculation has been altered–conditions that, when persistent, are linked to blood clots, heart attacks, and stroke.

Researchers had debated whether EMFs could induce rouleaux formation, but prior studies were performed in test tubes, criticized for stripping blood of its natural rhythms. Rob and his colleague Barbara Biebrich have brought the debate into the living body, where physiology unfolds in real time in a healthy human being. For the first time, the phenomenon was visible, undeniable, and alive.

Questions That Now Demand Answers

The implications of this discovery ripple outward. If five minutes of exposure can shift the behavior of blood cells in a healthy person in one exposed region of the body, what might hours of exposure do to other parts of the body that also can be regularly exposed? What about days, weeks, years–particularly in organ systems that depend on delicate microvascular flow?

What does this mean for the eye, with its fragile capillary networks? For the inner ear, with its electrically tuned hair cells? For the thyroid, resting in the very place where many people hold their phones? And what might be happening to children, whose thinner skulls and growing bodies absorb proportionally more radiation, whose tissues are still differentiating, whose lifelong exposure begins in infancy?

Rob does not sensationalize these questions. He presents them with the calm urgency of a clinician who understands that unanswered questions can be as dangerous as wrong answers.

The Good News–and the Unfinished Work

For now, the blood changes Rob observed are reversible. An hour or more after the exposure ends, the rouleaux dissipate, the cells drift apart, and circulation returns to baseline. But biological systems have limits. Reversibility under acute exposure says nothing about the effects of chronic or cumulative exposure–conditions under which the modern world increasingly operates. We do not yet know the point at which adaptive biological responses become maladaptive, or whether that point varies by age, health status, or environmental burden.

What Rob does know–and demonstrates with admirable restraint–is that this is a call for serious research, not fear. And that, tragically, is where society has faltered.

Federal agencies and industry leaders have not invested in the kind of rigorous, long-term studies needed to understand the full implications of continual EMF exposure. Instead, this responsibility has fallen to independent scientists and nonprofit organizations like Environmental Health Trust. That work continues only because thoughtful readers, concerned parents, visionary philanthropists, and public health advocates understand that knowledge is our greatest shield.

A New Frontier for Medicine

Rob's insights open the door to a new frontier in medical imaging: using ultrasound and other real-time technologies to observe how living tissues respond to electromagnetic forces. This frontier has the potential to revolutionize our understanding of environmental health.

Imagine being able to watch, in real time, how EMFs influence microcirculation in the eye, electrical pathways in the ear, or thyroid tissue during growth. Imagine mapping these effects in children, in pregnant women, in the elderly, in those with chronic illness. This is not science fiction. It is simply science waiting to be funded.

A Call for Wisdom

As a species, we have always advanced more quickly than we have understood the consequences of our own inventions. But we are also capable of great wisdom, especially when we listen—to evidence, to experience, to one another.

Rob Brown's book is a guide to that wisdom. It invites us to see the invisible—make the invisible visible—as the motto of Environmental Health Trust holds. To question with curiosity rather than fear. To acknowledge both the promise and the peril of the electromagnetic world we now inhabit. To recognize that children, pregnant women, and those facing illness deserve clarity, not uncertainty. To work with Rob and others at Environmental Health Trust to educate and motivate homes, schools, and workplaces in ways to reduce exposures.

Above all, it reminds us that we are electrical beings living in an electrical age. The forces that sustain us deserve our respect. The technologies we create deserve our scrutiny. And the future we build deserves our most careful attention.

An Invitation–and a Hope

It is my honor to call Dr. Brown a colleague and friend. His work, and the work of Environmental Health Trust, stands as a testament to what independent science can achieve when guided by integrity and a steady moral compass. Rob has written a bold and important book that expands greatly on efforts begun by Drs. Ronald Herberman and David Servan-Schreiber (both deceased) and I at the University of Pittsburgh Center for Environmental Oncology at the Cancer Institute nearly two decades ago. I am proud that our work and my books laid some of the foundations for Rob's efforts and thrilled to see the progress that has been made in understanding why and how we need to reduce exposures.

I highly recommend you read this book with an open mind and a listening heart. It will change the way you see the world–not by frightening you, but by helping you see the beauty, vulnerability, and resilience of the electrical life within us and the plants and animals around us.

Devra Lee Davis, Ph.D., MPH

INTRODUCTION

Like all of us, I am guided by the people, places, and events in my life, and, in my case, they have prepared me to share my experiences. Along the way, in addition to documenting a phenomenon I believe to be critically important; I have gained a deeper understanding of my surroundings. Ideas for this book have been steeping in the recesses of my mind for many years. After my father's passing in 2019, I felt an increased urgency to share this material. But my focus on completing the task stalled during the COVID-19 pandemic and all that ensued in its wake. So much continues to be discovered–or uncovered–while, at the same time, suspicion and distrust seem to be at an all-time high. As my own understanding of and research into the health effects of electromagnetic radiation have become more proficient, scientists and healthcare practitioners around the world have been performing research, writing scientific papers, and speaking at conferences expressing concern. Yet technology marches on. It's been a confusing time.

I've written blogs, been interviewed on various television programs, radio shows, and podcasts, and spoken to thousands of people over the years–in one-on-one conversations and at conferences–warning of the dangers of wireless technology. Despite my obvious passion for the subject, it has been difficult to reach people and get them to change their behavior. To tell people that one of their favorite possessions, their mobile phone, has been potentially damaging their health has been like trying to take a binky from a baby. Oftentimes, people listen and understand–until it means making lifestyle changes. Others don't want to listen, dismissing me with something like, "Thanks for the information, but . . ." As time goes by, however, I am hearing more and more people say they are interested and want to know more. I believe that's because most people intuitively know. They worry that the radiation emitted from the phones, Wi-Fi routers, and cell towers popping up all over the place may be affecting their health. We are a little afraid of the technology, yet we are uncannily attracted to its energy, like moths to a flame.

After making it through the worst of the COVID-19 pandemic, I ran out of enthusiasm for writing. Summer was fast approaching, and my teenage daughter was scheduled to fly to Uganda to volunteer for a nonprofit organization. Although I knew she was capable of navigating the logistics herself, I chose to accompany her to Amsterdam. Afterward, I planned to stay somewhere culturally vibrant to relax, regroup, and start writing again. I needed to be reinspired. Paris seemed like the perfect place to go, so I headed there after we parted at Schiphol.

I had come to Paris to write. Sounds about as clichéd as you can get, doesn't it? But I figured I could spend time sightseeing too. Paris is a wonderful city filled with exceptional art and culture. To not take advantage of its gems would be a mistake. I made a list of museums to visit in between my creative efforts. Upon my arrival, I immediately headed out to sightsee. I am usually good at reading maps, and this city wasn't too hard to figure out. However, while looking for my first museum, I got lost and found myself wandering down a collection of streets near the embassies. I knew the museum of art "moderne" was close, but the small map I had printed out was a little hard to read and I couldn't tell exactly where I was. After walking all over the small neighborhood, I caught a glimpse of a man down the street. I quickly headed in his direction and apologetically asked him for directions. He didn't speak English, but understood my dilemma as I pointed to the name of the museum on the map. He spoke to me in French and motioned with his hands. Embarrassed, I realized I was in the wrong neighborhood, folded up the map, nodded to him in appreciation, and hurried on my way.

I felt like I had been temporarily lost in a labyrinth. It was a disorienting experience. When I regained my senses and reached the museum, I breathed a sigh of relief. Then, I walked up the white staircase and into the first exhibit. I hadn't researched the museum, so I didn't know what exhibits were on display. But as a fan of modern and contemporary art, I figured I would enjoy the experience, regardless of which artists were featured.

The first room I walked into contained a mural by Raoul Dufy, a painter I was unfamiliar with. Complimentary headsets were available to provide a description of the artist, his work, and its impact. Before listening to the audio, I sat quietly on a small stool and gazed upon the floor-to-high-ceiling mural. It was a very colorful piece that covered the entire oval-shaped room. The center top of the mural depicted the Greek gods, beneath whom there were bolts of lightning. Then, my eyes began to focus on dozens of figures that filled the lower level of the piece. Next to each was a name: Gauss, Ohm, Faraday, Ampère, Volta, and many more. It took me a moment, but then I realized these were familiar names. I had been studying their contributions to the scientific world of physics for several years in preparation to write this book! These were the men who discovered electromagnetism, quantified electromagnetic radiation, and created the foundation for our technology. My mind clouded over for a moment, and I was filled with a sense of glee that tickled my entire body and lifted my spirit. An enormous smile filled my face as I thought, *Holy shit! This is an incredible manifestation!* I wanted to jump up and tell someone about this coincidence, but instead I sat and contemplated.

The real-world applications that derived from the genius of these men were illustrated in the mid-level of the piece. The creation of electricity, ships, power plants, railcars, buildings. Observing the progression of discovery to technology in this magnificent work of art was humbling. I thought of my concern that wireless technology was potentially dangerous and, at the same time, how the discovery of electromagnetism and our ability to create and manipulate this form of energy was extraordinary. Perhaps our current technology only needs tweaking to bring ourselves back on the right path. Is it really about how we wield this enormous power that is the greatest concern?

As I sat and stared at the massive painting, I thought of the connection between my desire to be inspired and my receiving this gift–a view of the entirety of the history of electricity and its founding scientists. Their discoveries were born from genius. My experience, a miracle of manifestation. Stumbling upon *The Spirit of Electricity* will be the moment I remember most about my trip to Paris.

Why am I qualified to present this material? My career as a medical doctor, interpreting diagnostic medical imaging studies, may seem like a far stretch from engineering cell phones to some. Yet during a radiologist's residency training, a full year of physics–including the physics associated with radiation and radiation biology–is part of the curriculum. Although a radiologist no longer operates the machinery that takes the picture in most parts of the world, we need to understand how the machines work and how the images are formed, how X-rays and gamma rays pass through the body but are also scattered and absorbed at the same time. We learn how electrical current and radiation intensity affect image quality.

When I began hearing that scientists around the world were concerned about negative effects from the radiation emitted by cell phones, Wi-Fi routers, and other sources, I was quick to learn what I could about the topic. Having a foundation in radiation physics certainly helped me grasp the concept that electromagnetic energy can affect living systems in ways that might not be readily apparent. But there was a lot to integrate. As will be covered, the radiation produced by X-ray equipment is similar to that produced by cell phones, but also different. Despite my physics education, I didn't have a firm understanding of the components of the electromagnetic spectrum. The relationship between electromagnetic fields (EMFs), radiofrequency (RF) radiation, and X-rays felt abstract and was of no consequence to me during the early part of my career. When smartphones came on the market, I was enamored like everyone else with the ability to carry all of the world's accumulated knowledge in my pocket. Yet, I must admit that I, like many of you, was unconsciously suspicious of its safety.

I am not afraid of radiation, but I do respect it. Radiation is a form of energy that comes in a complex diversity as vast as that of matter. The enormous variations in the characteristics of radiation determine its effect on matter. Einstein showed us how the two are intricately related with his formula $E = mc^2$. Think about the differences between solids, liquids, and gases–each a state of matter with a different energy content. If you simplify the model and think about the differences between ice, water, and water vapor–all states of matter with the same chemical composition–you can imagine how differently they can interact with living beings. For example, you can't inhale ice, but you can inhale steam. You are able to stand on ice, but you probably can't stand on water. These examples may seem silly, but I'm making a point that energy quality and content within matter determines how living beings interact with their environment, especially the electromagnetic spectrum.

Most of my colleagues are unconcerned about cell phone radiation. They acknowledge that various diseases have become much more prevalent in recent years, but the science is considered fringe for most physicians because it inhabits the obscure intersection of biology, physics, engineering, and medicine.

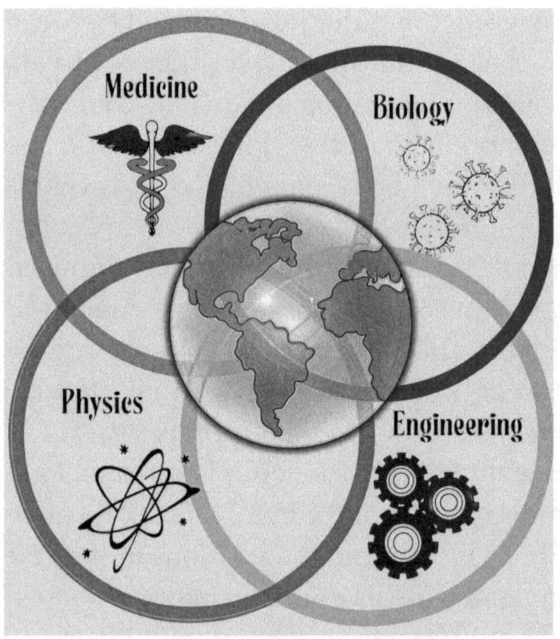

A multidisciplinary science

Most physicians know very little about physics and even less about engineering. Physicists don't understand medicine or biology. Engineers have learned some physics, but haven't been trained in the intricacies of medicine or biology. As such, we are dealing with technology that very few fully grasp for it involves four vast scientific fields.

Writing and lecturing about EMFs and microwave (MW) radiation is difficult, in part because it is intangible. This energy cannot be heard or felt. It has no odor or fragrance.

For most people, the effects of chronic exposure are untraceable, manifesting insidiously over unknown periods of time in a variety of conditions. It is the slow depletion of energy and zest for life that can be the hardest to trace. I find creating analogies that resonate to be an effective method for communicating this complex information. Everyday examples can help people understand and relate to complex ethereal concepts.

While eating breakfast at a small guesthouse on Lago Garda in Northern Italy, my two fellow travelers, my sisters, commented on the intense color of their scrambled eggs and my yolk as I carefully scooped the top off a soft-boiled egg. The color of the egg yolks *was* remarkable. They were a rich, dark yellow, bordering on orange. And they were delicious. I remarked about how these eggs were obviously produced by healthy, pasture-raised chickens. My sisters didn't understand why I made that assumption. As I explained, I realized it would be a perfect analogy with which to start this book and illustrate how incessant exposure to electromagnetic radiation (EMR) can cause depletion over time.

Most of us buy eggs from the grocery store and don't think too much more about them. Eggs are presorted and packaged according to their color and size. Eggs may be conventional, meaning the laying chickens are raised in factory hen houses, or they may be marketed as free range, pasture raised, organic, or organic free range. The tastiest eggs are produced by chickens who have exposure to grass, insects, weeds, and . . . the sun.

If you have ever raised backyard chickens, you know that the frequency of egg production depends on the season. The color intensity of the egg yolk is also variable. Typically, yolks produced from backyard chickens, fed a diet of untreated grasses and insects, are much richer and darker than store-bought yolks.

When a hen lives its life in sync with the earth's rotation and wobble, her egg production is maximized during the longer days of spring and summer and drops off during the shorter days of late autumn and winter. This cycle gives the hen time each year to regroup and replenish her nutritional status. It is biologically important for her eggs to be filled with proper nutrition so, if the egg is fertilized, it will have the necessary nutrients to produce a healthy chick. Eggs early in the season are smaller with firmer shells. As the hens increase production during the season, egg size increases, shells become thinner, and yolks become less intense in color as the chicken's nutritional status is slowly depleted.

Factory chickens aren't raised with natural light and dark cycles. In order to maximize egg production, these hens are exposed to prolonged hours of light all year, as if it's always summer. Floodlights trick the chickens into continuing to produce eggs during the winter. This method is great for increasing annual egg production, but doesn't allow time for the chicken to regroup and reaccumulate nutrients. Those of us used to buying eggs with light-yellow yolks from the grocery store are used to eating less nutritious, less tasty eggs produced by depleted hens. How is this story related to cell phones? As we will discuss, the radiation produced by cell phones and light share something in common. They are both forms of radiation, specifically electromagnetic radiation (EMR), a term that will become very familiar as you progress through this book.

It's funny how we get used to our environment and don't perceive stressful elements that are constantly there. This doesn't mean they aren't causing us any damage; it just means that our brain has learned to ignore these inputs, so we are no longer conscious of them. But, when toxic elements are removed and our health status changes, we can appreciate the change as we reconnect with how we used to feel.

Working to educate the somnolent populace has become my mission. I speak to people about EMR every day. It comes up in daily conversation because everywhere we look there are sources of human-generated "anthropogenic" EMR, both indoors and outside. Governments have green-lighted deployment of this technology, which is exposing us to radiation every day, and some of us, twenty-four hours a day. As technology advances, the stakes are getting higher. People everywhere are developing symptoms. Yet, they don't connect their health problems to this radiation. In a very real sense, our world has become enveloped by electropollution, an invisible assailant that is becoming inescapable.

Thousands of peer-reviewed scientific articles have detailed the damaging health effects EMR can have on living beings. Research has been performed on cell lines, animals, and people. Although study results presented to the lay public and policymakers have been determined to be ambiguous, I am convinced from what I have read and experienced in my own life that the biological effects caused by RF radiation are real. Given the increasingly incessant exposure to which we are all being subjected, I fear we are on the precipice of a major decline in public health. I have written this book to explain the many ways in which anthropomorphic EMR can damage human health. By learning a bit about this technology, its bioeffects, and how to reduce unnecessary exposure, I believe we can enjoy the benefits of technology without sacrificing our health.

Beyond that, why should you read this book? I truly believe our societal exploration into the world of energy manipulation has provided us with an opportunity to understand more deeply who and what we are. At our fundamental core, we are energy. Learning how we interact with energy in our environment may increase consciousness. Although not yet defined scientifically, we all have stories that illustrate the apparent energetic connections we share–with one another as well as with the Earth. The more we can understand these relationships, the stronger our sense of self and our relationships can be.

This is a book you should read from the beginning and work your way through, chapter by chapter. My method is to first present technology and then its effect on biology. The physics described in the early chapters is important to understand before engaging later chapters. My descriptions at times may seem overly simplistic in the eyes of engineers, physicists, and other professionals—and they are in some cases. But I think that is necessary to help laypeople to understand what's at stake.

Although I have attempted to use language for a nonscientific audience, many concepts require the use of scientific terminology. A glossary has been included to make the definition of terms readily accessible. The science behind EMR is expansive, but it is not necessary to understand it all. If you approach this book with an open mind and a willingness to learn, you will be amazed at both the genius of technology and the incredibly complex elegance of biology, as well as the delicate interface between the two.

CHAPTER 1

VISIBLE LIGHT

We live in a world filled with both natural and human-generated sources of energy. Terms like *microwave* and *Wi-Fi* are familiar, but do you understand what they actually represent? Both are invisible forms of radiant energy. Like visible light shining from the sun, radiation emanates and diverges in all directions from a central point of creation.

Energy is abstract. With the exception of visible light, we can't see pure energy, but we can appreciate its effect on objects or interpret its character with our other senses. We can observe the effect wind energy has on a body of water and watch waves propagate and eventually crash on the shore. We can hear sound energy and feel the radiant heat of a sauna.

Early on, the nature and properties of light were not understood. It may seem obvious to us today that light arises from objects that emit this form of energy–such as the sun, a candle, or an LED screen–but human understanding of light has evolved from an initial belief that light was created by the human eye. This was the accepted explanation for light until 965 CE, when Alhazen, a Persian scientist, proved light was created by objects outside our bodies.

In the 1600s, Newton originally explained light as being composed of tiny particles. His theory was later supported by Einstein's discovery that light travels in discrete packets of energy, called *quanta* or *photons*. A particle theory of light explains some of light's characteristics, such as reflection, but not others. In the late seventeenth century, the Dutch physicist Christiaan Huygens performed an experiment that indicated the characteristics of light were more complex and could not be explained by particles alone. Huygens shined a light through a narrow slit and onto a distant screen, creating a pattern of alternating light and dark bands on the screen. The brightest, widest band in the middle was referred to as the *central maximum*. On either side of the central maximum, multiple symmetric, less intense secondary bands were observed. Huygens recognized this as a diffraction pattern, one that is characteristic of waves, not particles. Huygens's principle asserts that each part of the wave front creates additional wavelets that are in phase with one another at the slit aperture, but beyond the opening, the waves intersect one another, much like the overlapping of ripples created by throwing multiple pebbles into a still pond. The result is an interference pattern on the distant screen (Figure 1.1). His experiments also showed that light waves were not unidirectional, but rather a form of radiation, emanating in all directions from their source.

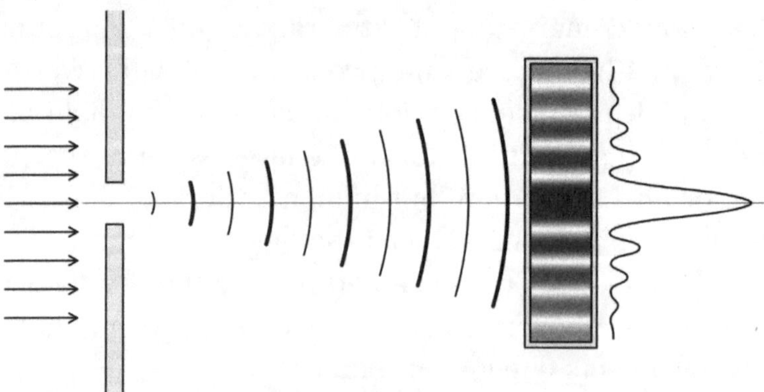

Figure 1.1. Single-slit diffraction

In 1801, Thomas Young performed a similar experiment to Huygens's in which he shined a light source through two pinholes, each cut out from the same piece of paper. This double-slit experiment also generated an interference pattern, a sequence of alternating light and dark bands on a screen, which was again recognized to be characteristic of waves (Figure 1.2). Young was also the first to propose that light travels in a transverse wave, as opposed to a longitudinal wave, which enables electromagnetic waves to be polarized.

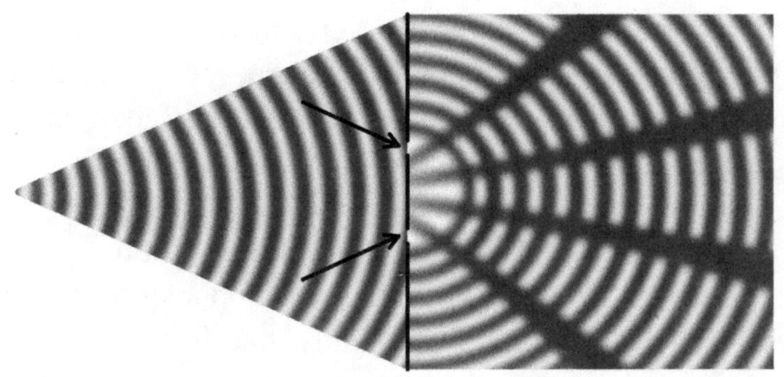

Figure 1.2. Double-slit experiment

What's the difference between transverse and longitudinal waves, and what does it mean to be polarized? To demonstrate a longitudinal wave, imagine handling a child's Slinky toy. If you were to hold the ends of a stretched-out Slinky relatively stationary from side to side, and then quickly push one end of the Slinky in and then out again, the compression of the coils would travel from one end of the spring to the other, in the same direction the wave is traveling. This movement is a representation of a longitudinal wave's propagation. The wave transmits without any up-and-down or side-to-side motion. Sound travels in longitudinal waves.

Transverse waves propagate differently. If we watch a wave travel the surface from one end of a water tank to the other, we see the water rise and fall, moving up and down. The shift in the molecules' position, then, is in a direction *perpendicular* to the direction that the wave travels. This is a *transverse* wave. If a wave, as in this case, only has a vertical component and no horizontal component, it is *polarized*, with all of the vibration happening in a single plane. Electromagnetic waves from a natural light source are not polarized because they vibrate in every direction perpendicular to the direction of travel. Mathematically, each wave's vibration can be described by a horizontal component (x) and a vertical component (y), just like a vector in a coordinate plane. Natural sources of light can become polarized by reflecting off some surfaces, creating glare which can negatively affect our vision. Polarizing lenses make the best sunglasses because they eliminate glare.

The discovery and subsequent confirmation that light is composed of waves carrying discrete packets of energy, referred to as a *wave-particle duality*, had huge implications because waves have specific properties and characteristics. It is important to understand the properties of waves because this foundation will apply when learning how waves impact physical matter. If you are unfamiliar with the properties of waves such as frequency, penetration, and resonance, please refer to appendix A.

The sun produces and emits particles and ions, as well as the gamut of electromagnetic frequencies spanning from extremely low-frequency radio waves to the highest-frequency gamma waves. We are mostly protected from high-energy particles emitted by the sun by their deflection off the Earth's magnetosphere. At the poles, where the magnetosphere is weakest, some ions can penetrate through and interact with atoms in the Earth's atmosphere, creating the brilliant light spectacle known as the aurora borealis, or northern lights.

What we experience as sunlight is a bandwidth of electromagnetic energy that includes visible light and portions of the infrared and ultraviolet (UVA and some UVB) spectra below and above visible light frequencies. A subset of lower-frequency radio waves can also pass through the atmosphere and reach the Earth's surface. But most other frequencies of EMR are blocked by our thin atmosphere and ozone layer. This is fortunate for life on Earth for frequencies higher than UV-B–including UV-C, X-rays, and gamma rays–carry so much energy that they can cause direct ionization of atoms. This means that when these energy waves are absorbed, they can blast an electron out of its atom's orbit, turning the atom into an unstable ion. This ion then looks to replace the missing electron from any nearby atom. When this happens in a living being, it can wreak havoc and do serious harm, including causing DNA damage or cancer. It is for this reason that the production of *ionizing radiation* is heavily regulated by governmental agencies, such as the FDA in the United States.

Life on earth has evolved with this filtered bandwidth of EMR and has developed physiological adaptations to detect or respond to these wavelengths. Plants harness energy from the visible light spectrum and trap it into physical form by creating chemical bonds and constructing sugars and carbohydrate molecules. This process is known as photosynthesis.

Photosynthesis

The sequence of electrochemical reactions that occurs during photosynthesis has enabled life to exist and thrive on Earth. The energy in visible light photons isn't huge; in fact, it averages around 12 eV (electron volts), slightly less than the 13.6 eV needed to cause ionization of hydrogen, the creation of a charged hydrogen atom by removal of an electron. The absorption of light energy during photosynthesis "excites" electrons. The energized electron moves farther away from the atom's nucleus into an outer orbital with a higher energy state. It is the movement of this electron back to its home orbital that releases the energy needed to cause biochemical reactions. Yes, 12 eV seems like a very small amount of energy but what may be even more surprising is that experiments have shown that chlorophyll strongly absorbs red light frequencies, and red light only has an energy of 1.8 eV. How could it be that photosynthesis occurs from a photon with only 1.8 eV?

Photosynthesis can't happen just anywhere. It can only occur within the chlorophyll molecule, which is housed in a specific plant cell organelle called the chloroplast. Chloroplasts are rich in metals, including iron, copper, and to a lesser extent, manganese.[1] The chlorophyll molecule is a *porphyrin,* meaning it is a complex molecule that forms a cage-like three-dimensional structure with a central sequestered metal ion. In chlorophyll, the central metal ion is magnesium (Mg). It is widely known that electrons move more easily in metals than in non-metals. The presence of Mg likely decreases the binding affinity of nearby electrons for their respective nuclei, reducing the energy required to bump those electrons into higher orbitals when absorbed from a light photon. The excited electron is then picked off by a molecule called a *mobile electron carrier,* effectively causing ionization, also known as *oxidation*, of the chlorophyll molecule.

This is not the direct ionization that physicists mean when they talk about "ionizing radiation." The electron isn't shot out of the atom completely by the photon, but rather moved into a higher energy state from which the electron can then be carried away. The net result, though, from both processes is ionization.

During photosynthesis the excited electron's energy is controlled and used to generate energy molecules called *adenosine triphosphate* (ATP). Meanwhile, the chlorophyll molecule regains its lost electrons from nearby water molecules. The water molecules then become oxidized, generating oxygen and hydrogen ions. The chloroplast uses its generated ATP to create sugar through a series of chemical reactions called the Calvin cycle.

I am going into detail about photosynthesis because it is an omnipresent example of how nonionizing radiation can cause profound physiological effects in living organisms. The presence of a precisely chosen and positioned metal ion can create a microenvironment in which atoms can become ionized more easily and at much lower frequencies than otherwise expected. To reiterate, in the chlorophyll molecule, a Mg atom enables ionization by lower frequency light, with longer wavelengths, in the visible spectrum.

Visual System

Our visual sensory system, as well as those of many other animals, has developed receptors to detect the narrow bandwidth of frequency we call the visible light spectrum. We humans are limited to seeing this narrow spectrum, but that is not true for all animals. Some insects, including mosquitoes, beetles, and bedbugs, can see into the infrared bandwidth. Vampire bats, frogs, fish, and some snakes can also see infrared energy. Ultraviolet wavelengths can be sensed by bees, sockeye salmon, and apparently reindeer! Interestingly, the miraculous butterfly, the spiritual symbol of transformation, has the ability to detect the widest bandwidth of EMR of all life-forms.

The overall design of the human visual system is extraordinary. Our eyes are covered by a white sclera. Because white surfaces reflect all colors of the visible light spectrum, perhaps the sclera acts as a shield, effectively blocking the entry of light into the eye, except for those rays that pass through the pupil. One might think that just having a barrier of tissue, i.e., the outer surface of the eyeball and overlying soft tissue, would prevent the entry of unwanted light, but consider that even with eyelids closed, we can still sense light because EMR passes through the skin of the eyelid and through the pupil.

The pupil's size varies depending on contraction of the iris, the colored band encircling the black pupil. The iris contracts or dilates in response to radiation levels. In bright daylight when there is a higher concentration of blue light, the iris constricts, allowing less light into the eye. During sunset or daybreak, the dimmer light and orange color of the sun result in a wider aperture of the iris. Longer red wavelengths are safer for the retina than higher-energy blue wavelengths. A responsive iris helps to protect the internal structures of the eye from excessive EMR exposure. Once light passes through the pupil, it is refracted through a biconvex lens designed to focus the light onto the retina, the lining of the back wall of the eye.

The retina contains specialized cells called rods and cones, which absorb the packets of light energy we call photons and convert that energy into electrochemical signals. In this process, nonionizing radiation within the visible light spectrum causes specific molecules in the retina to change shape and generate electrochemical signals, which are then transmitted to the brain. Rods are extremely sensitive to light and produce a chemical reaction to even a single photon of light. Photoreceptive chemicals within cones allow these cells to differentiate light's frequency, and therefore its color, information that is then conveyed through the optic nerve to the brain. The visual cortex in the brain receives these electrochemical signals and interprets the image.

We observe a leaf as green because the green wavelengths of EMR are reflected by the chlorophyll molecules while the other wavelengths in the visible light spectrum are absorbed. It is the reflection of specific wavelengths from objects that determine their color. A red bird, for example, is observed as red because its feathers reflect red wavelengths of the visible light spectrum while absorbing all other wavelengths.

Visible light accounts for only a sliver of the electromagnetic spectrum, yet it is profoundly important for our understanding of the world. If we didn't have a retina with cells that were able to absorb this bandwidth of radiation, we would not see "light." These frequencies would be as undetectable as most other frequencies of EMR. It is important to recognize that there is no fundamental difference between frequencies within the visible light spectrum and other frequencies in the electromagnetic spectrum.

Circadian Rhythm

Almost all living beings on Earth have evolved in response to cyclical EMR exposure. Our periods of light and darkness are the product of the sun's constant radiation from a relatively stationary position and Earth's continual rotation and its tilt. Even bacteria[2] and some viruses respond to cycles of light and dark![3] Diurnal organisms are designed to be active during the daytime when the sun is shining in the sky, and quiet at night when skies are dark. Nocturnal animals, on the other hand, are active at night and slumber during the day. Regardless of whether you are a day person or a night owl, the daily recurring pattern of daylight and darkness affects your behavior and defines what we call a "day." Our sleep/wake cycle, in addition to many other bodily functions, is entrained to the periodic variations of light and darkness. These processes collectively make up what is referred to as our *circadian rhythm*.

The German zoologists Hans Kalmus and Erwin Bünning originally discovered circadian rhythms in fruit flies in 1935, but a better understanding of life's complex and intricate relationship between light/dark cycles on Earth has evolved over time. Scientists continue to discover new biological functions that fluctuate with the timing of our exposure to–and relief from–the sun's electromagnetic radiation.

In humans and in most vertebrates, circadian rhythms are controlled by a small structure in the brain called the pineal gland. Although insignificant in size, the pineal gland has been regarded as a spiritual beacon for centuries. In fact, Rene Descartes (1596-1650) considered the pineal gland to be the seat of the soul. Many spiritually oriented cultures around the globe refer to the pineal gland as a third eye. This makes some biological sense because, like an eye, the pineal gland's function is governed by the presence or absence of light. I marvel at the fact that ancient cultures knew this.

The Pineal Gland

The outside world signals the pineal gland indirectly via the eyes. Neuronal pathways connect the retinas to a portion of the brain called the hypothalamus, which in turn communicates with the pineal gland. These pathways, referred to as the *retinohypothalamic tract*, only transmit information about the presence or absence of light. This distinct pathway is separate from the optic nerves, which carry visual information.

At the end of a prolonged period of brightness, as would naturally occur during daytime, the dimming of ambient light at dusk is detected by specialized cells in the retina, which send signals to the *suprachiasmatic nucleus* (SCN) in the hypothalamus, alerting the brain that it's time to prepare for sleep. The SCN functions as our master internal clock. When it receives the message that the surrounding environment is dark, it relays this information via a neural pathway to the pineal gland. In response, the pineal gland produces and releases the sleep hormone melatonin.

The amount of melatonin produced and released by the pineal gland is proportional to the length of the period of darkness.[4] Therefore, in a natural world, longer nights will generate larger amounts of melatonin. In a world without artificial light, in the northern latitudes, melatonin production would be maximal during the winter months and minimal during mid-summer months, when days are shortest. The inverse is true for the Southern Hemisphere. The natural seasonal variation in melatonin production connects us biochemically to the Earth's rotation, orbit, and tilt, i.e., the angle of the Earth's axis that gives us seasons.

You might imagine that during a full moon, because the sky doesn't completely darken, melatonin production would be less than during nights with new moons or crescent moons. Perhaps this is why moods, behavior, and exacerbations of disease have been observed during a full moon both in humans and in animals.[5, 6] In the past, this association has been attributed to folklore, but as more scientific research emerges, the correlation of a full moon to the exacerbation of disease is being proven to the scientific community.[7] It may be only anecdotal, but I have always dreaded being on call during a full moon.

Melatonin

This biochemical adaptation helps ensure that we, and other life-forms on Earth, function optimally and have downtime to repair our bodily systems. Sufficient melatonin production is crucial for good health. Biochemically, the pineal gland converts the amino acid tryptophan to serotonin, which is then converted to melatonin. It is therefore important to have sufficient tryptophan in one's diet. While melatonin supplements have become popular for inducing sleep, melatonin provides a wide assortment of other health benefits that directly affect our state of well-being. Hundreds of scientific research studies have documented melatonin's numerous contributions to the immune system and to maintaining good health. This hormone has antioxidant and anti-inflammatory effects and is extremely important in tumor suppression.[8]

One of the repercussions of using lamps and other sources of artificial illumination at night is that they unnaturally increase the length of our day; at least that is how our retinohypothalamic tract sees it. Clicking on a lamp or turning on the TV at night, instead of winding down for sleep when the sun goes down, causes the brain to stay alert and melatonin levels to drop. Outdoor lights, including streetlamps, car headlights, and illuminated buildings and landscapes, affect not only people, but the animals and insects that gaze upon these sources of EMR.[9] I used to live in Florida, where road signs along the beach warned about the use of streetlights during turtle nesting season. This concern was based on the observation that when newly hatched sea turtles see streetlights, they instinctually crawl toward them, thinking they are moving toward the ocean water. As a result, the baby turtles risk being run over and killed by cars. Over the years, I have come to realize it is not only the sea turtle that is affected by streetlamps. We all are. I make it a point to turn off all outdoor lights on my property. Although many have suggested over the years that I should spend money on exterior illumination, I find that driving down my driveway at night into the darkness is comforting. Deer, raccoons, and other animals seeking refuge on the property do too.

Most of us do not experience darkness at night. And as a result, our circadian patterns have significantly weakened. To add fuel to the fire, looking into the light-emitting diode (LED) screen display of a smartphone, tablet, or laptop is even more distorting to the natural circadian rhythm because these displays have a higher percentage of blue light wavelengths than ordinary incandescent bulbs. In adolescents, melatonin secretion can be diminished by up to 38 percent if the teenager is exposed to two hours of LED screen illumination before bedtime.[10] Some say we are now facing global systemic dysfunction.[11] I often wonder if this is why we have an annual cold and flu "season." Our immune systems should be in balance with the environment and fight off infection even in the dead of winter.

Instead of figuring out why the population gets sick, governments, health professionals, and businesses around the world focus on the "solution" of selling vaccines and medications.

Light Bulbs

It is possible to modify your indoor lighting to be more conducive to maintaining circadian rhythms. The color of the bulbs you choose for each room can either relax or stimulate the brain. But making your choices can be a bit overwhelming, particularly if you go into a store that sells only bulbs! The variety of colors and technology is immense. In general, bluish light will increase concentration and improve visual acuity, whereas warm reddish light promotes relaxation. This makes sense when you consider our perceived color of sunlight during different times of the day. The color-rendering index is a scale from 0-100 that rates an artificial light's ability to accurately represent colors as compared to daylight.

When choosing light bulbs for your home, try to select those with a high color-rendering index that simulate the most natural conditions possible. Importantly, also consider a bulb's color "temperature" when choosing bulbs. Bulbs with longer wavelengths produce a lower temperature (< 3,000 Kelvin) and a warmer, yellowish glow and are therefore more suitable for the bedroom and rooms for relaxation. Paradoxically, shorter wavelength light bulbs are labeled "cooler" even though they produce higher temperatures. These bulbs are typically between 3,000 and 4,500 K and are better suited for areas requiring more energy or focus, such as the office, study, kitchen, or playroom. Bulbs with color temperatures above 4,500 K produce a bright light, akin to natural daylight. Just in case you were wondering how a light bulb could produce such a high temperature, they don't. This scale is based on the temperature that a "black body emitter" would have to be brought to in order to produce the specific color of light.

Perhaps one of the most important additions to my home over the past ten years has been a sunrise alarm clock. This device initially illuminates with a dim reddish glow that slowly intensifies and changes to a more whitish color, simulating sunrise. The brain can see the illumination through the eyelids and stops producing melatonin. It allows one to wake up slowly and more naturally, without the stress of being jolted awake by a noisy alarm. As I start my workday at 4 a.m., this lamp has been extremely helpful.

If you occasionally need to get up in the middle of the night, I recommend placing a Himalayan salt lamp in the bathroom or in the hallway, wherever it is that you need to go to while it is dark in the house. These salt lamps produce a soft reddish glow, providing enough light for you to see where you are going without completely waking you. Although I have not read any scientific literature on their effectiveness, from my experience, I believe they are less disruptive to the circadian rhythm, and therefore a healthier alternative than turning on an overhead light, particularly if that light is a fluorescent light, daylighting bulb, or compact fluorescent light bulb (CFL).

In a move to improve efficiency, vendors have created CFLs and LEDs. CFLs are particularly problematic. Whether they are single or double encapsulated, should they break, they need to be cleaned up with a specialized technique to avoid mercury exposure. CFLs also produce ultraviolet radiation, which can be damaging to the skin, particularly in those more prone to skin damage from EMR. If you have chosen CFL bulbs in your home, make sure they are not used for close-up work, but rather remain at least several feet away from you while they are turned on. LED bulbs are also more efficient at converting electrical current into light than standard incandescent light bulbs. Unfortunately, both LEDs and CFLs produce dirty electricity, a phenomenon that will be discussed in chapter 4. Although it is becoming more difficult to achieve,

I would recommend lighting your home with ordinary incandescent bulbs.

Light bulbs are only one source of illumination we are exposed to at night. Screen time at night not only affects circadian rhythm and melatonin production, it can also damage structures of the eye itself. Our eyes have been designed to interpret the reflection of light off objects. Our eyes are not designed to look directly at a source of illumination. Staring into an LED screen can damage the vitreous humor, the fluid that fills the back chamber of the eye,[12] and the retina. Macular degeneration can also occur from too much screen time.[13] Other potential complications include headaches, ocular surface disease (such as dry eyes), blurry vision, cataracts, and glaucoma.[14] Thankfully, inventors have developed blue light-reflecting glasses to ameliorate some of these effects.[15] I initially purchased these glasses for my family early in the COVID pandemic. I found they reduce eye strain and fatigue.

Inventors have also developed backlit screens that emit lower levels of short wavelength radiation, i.e., blue light, reducing the risk for retinal damage.[16] E-readers, called "e-ink," which diffuse the light across the screen, are also an improvement to LED screens. The small amount of research done on these screens suggest they do reduce eye strain[17] but still disrupt melatonin production and release.[18] If you want to improve your melatonin production, try reading a print book by lamplight with a warm or reddish incandescent glow before heading to bed, rather than reading from an LED screen or an e-reader.

Light and Skin

We naturally think first about light and its effect on the eyes because they are the specific sense organs our body has developed to interpret electromagnetic energy in our surroundings. But the largest organ of the body, the skin, provides an interface and a boundary protecting us from excess visible light and ultraviolet radiation.

Skin blocks the penetration of light energy and some other frequencies of EMR by the dispersion of a molecule called melanin. Melanin is an extremely important and ubiquitous compound that is able to absorb and diffuse EMR; it's particularly effective at protecting the body from ultraviolet radiation. Both plants and animals utilize melanin to protect themselves from excessive solar radiation. Melanin has other functions too, but its protective role is crucial for our well-being. Those whose biological ancestors came from the equatorial region have more melanin in their skin because sunlight is much more intense with a greater concentration of blue-light frequencies in this area. Over many generations, the skin of people living around the equator has adapted to the frequent and intense solar radiation by producing more melanin to better protect the body. On the other hand, those ancestors who were from latitudes farther from the equator were exposed to much less direct sunlight and didn't require as much protection from melanin. Therefore, their bodies did not adapt in the same way. Today, people with different ancestry and different concentrations of melanin in their skin live everywhere. For those with a Scandinavian background, living in Florida may be problematic, as their skin is not designed to protect their body from more intense blue wavelengths of radiation. Conversely, those with whose ancestors came from equatorial Africa may find they cannot absorb enough sunlight to produce adequate vitamin D if they live in places like Canada.

Vitamin D

The absorption of EMR by skin also results in the production of an important hormone called vitamin D_3, a subtype of vitamin D. During summer, when short nights result in less melatonin production, long days of sunlight enable our bodies to naturally produce greater amounts of vitamin D_3. Like melatonin, vitamin D is an important antioxidant and critical for immune health. Vitamin D and melatonin sit on opposite sides of a light seesaw. When one side goes up, the other goes down, and vice versa. (Figure 1.3) These two molecules do not perform the exact same biological actions, but their reciprocal relationship to light is likely by design and not coincidence, keeping the immune system fortified during both winter and summer.[19]

Figure 1.3. Reciprocal relationship of vitamin D and melatonin

Vitamin D production occurs in several steps. First, high-frequency, nonionizing electromagnetic radiation, specifically UVB rays, penetrate the superficial layer of skin, called the *epidermis*. These UVB rays cause a chemical reaction (*photolysis*) of a molecule called *7-dehydrocholesterol* (7-DHC), turning it into *previtamin D3*. Previtamin D_3 will then either reconfigure into an active form, vitamin D_3, or again photolyze from UVB rays into one of two nonactive forms, *lymisterol* and *tachysterol*.[20] As with most bodily systems, a feedback loop prevents the overproduction of this potent hormone. When the body has enough vitamin D_3 on board, additional skin exposure to light will result in a natural diversion to synthesize the nonactive forms. This safety valve is subverted when one takes vitamin D_3 supplements.

Many of us do not get enough daily sunlight exposure. Working days are spent indoors. When we do make it outside, full-spectrum sunscreens block our skin's exposure to UVA and UVB rays. Sunscreens have helped reduce the incidence of melanoma, basal cell cancer, and squamous cell carcinoma.[21, 22] They also prevent premature aging of the skin and loss of elasticity. Given these important health impacts, dermatologists recommend wearing sunscreen all the time while outdoors. But an unintended consequence is that our natural production of vitamin D_3 has taken a significant hit.

It was once thought that vitamin D was only important because it prevented rickets, an easily recognized condition of abnormal skeletal development on X-rays. Vitamin D was added to milk as a supplement for children, but adults who didn't drink milk were not receiving this boost. It is now recognized that vitamin D's effect on skeletal growth is by no means its only role in maintaining homeostasis. Nearly every tissue in the body—including the brain, heart, muscles, immune system, and skin—are affected by vitamin D. Deficiency can lead to cancer, autoimmune diseases, cardiovascular disease, and neurologic disorders. A recent study showed a 22 percent decreased incidence in autoimmune diseases such as rheumatoid arthritis, psoriasis, and Grave's disease in patients who supplemented their daily intake of vitamin D by 2,000 international units (IU) over an average time frame of 5.3 years.[23] Inuits and others living near the Arctic Circle have low levels of vitamin D, especially during winter, when the sun barely rises above the horizon. However, increased dietary intake of salmon provides their people with high concentrations of vitamin D. Interestingly, the Inuits' digestive tract evolved to absorb more vitamin D through their food. They have also developed a more efficient receptor binding, which allows their seemingly low concentrations of vitamin D to be sufficient to fulfill their physiological needs.

Supplementation

Although supplementation can be very helpful and can prevent disease, vitamin D is fat soluble and too much can be toxic. Dosing should ideally be titrated, i.e., tailored to the individual and based on a person's size, habits, diet, and lifestyle. Vitamin D supplementation should be monitored by a physician who can order periodic blood tests to ensure vitamin D levels are sufficient and not too high. Importantly, vitamin D_3 can be stored in fat and remain available for the body's use, effectively prolonging its half-life to two months. Half-life is the length of time it takes for a substance, in this case Vitamin D_3, to diminish or decay by one-half its original amount. In the northern latitudes (or southern in the Southern Hemisphere), we should ideally be able to store up enough vitamin D in the fall to carry us through the darkest months of winter.

Melatonin supplements have also become extremely popular. They are regarded as safe, and few people experience significant side effects. However, I have never personally used them, and I am hesitant to recommend them. Unlike vitamin D production, the physiological timing of endogenous melatonin release is closely tied to one's circadian rhythm. I am not sure that a melatonin supplement would be ingested and absorbed at the time and concentration to properly support endogenous production. In my opinion, it is better to strengthen your circadian rhythm to optimize your body's natural melatonin production, so it is released when physiologically appropriate.

Here are several recommendations for optimizing your melatonin production.

- Pick a bedtime and stick with it. Your sleep cycle is an integral part of your physiology, and consistency will result in a more predictable release of melatonin.

- Avoid exposure to LED lights and LED screens for thirty minutes before bedtime. If you view screens at night, wear blue light-reflecting glasses and change screen settings to warmer hues.
- Remove all sources of illumination from your bedroom. These include tiny LED lights that illuminate electronic equipment. Either unplug these devices or cover them with black tape. Place blackout shades on your bedroom windows to limit ambient light coming in from outdoor lights. Dim alarm clock displays. If you still can see in your bedroom at night, purchase a sleep mask and wear it every night.
- Reduce EMF levels in your bedroom. In the remainder of the book, we will cover different types of EMF and their sources. But, for now, simply unplug your Wi-Fi router at night and place cell phones in airplane mode or turn them off altogether before you head to bed.

Light and Healthcare

Light can affect mood and is often used as a form of therapy. Seasonal affective disorder is perhaps the most widely recognized condition that can occur during the winter months, when exposure to natural sunlight is limited because days are short and nights are long. The depression people experience can be ameliorated by exposure to full-spectrum light, simulating sunlight. Treatments can improve focus, elevate mood, and normalize sleep patterns. The timing and duration of exposure are important to maintain proper circadian rhythms. Treating someone with full-spectrum light addresses the cause of the problem and could be a better option than prescribing them antidepressants.

Some skin conditions can be treated by exposure to frequency-specific LED lights. Red light treatments have been shown to stimulate collagen growth and are effective at repairing aging skin and reducing inflammation. Red light wavelengths are at the longer end of the visible light spectrum (633 nm) and therefore penetrate more deeply into the skin than blue light. A study by Ruggiero in 2016 showed that red light frequencies can penetrate skin to a depth of 5 mm.[24] Blue light LED with a wavelength of 415 nm transmits a higher energy, but only penetrates to a depth of 2 mm. Blue light therapy has been shown to be effective at killing bacteria and diminishing the activity of sebaceous glands, thereby improving acne, rosacea, and psoriasis.[25]

Lasers

Once looked at as a technology in search of an application, lasers have found plenty of them in recent years. Although many of us only experience lasers at the grocery checkout, I am including a discussion here because they illustrate how powerful EMR can be when it is focused and manipulated with precision. The word *laser* is an acronym that stands for **l**ight **a**mplification from **s**timulated **e**mission of **r**adiation. A laser is created by generating a single frequency of EMR between the infrared and ultraviolet spectra, including visible light, and then intensifying that beam by adding additional coherent waves, meaning the peaks and troughs are in sync with each other. The additive effect of the superimposed waves significantly increases the intensity or amplitude. The light is then highly collimated, or focused, so the energy does not disperse. The result is a beam of light that is much more intense and can stay focused for a much greater distance than light produced, say, by a flashlight.

Lasers are used in medicine for procedures as diverse as LASIK (corrective surgery for nearsightedness), tumor shrinkage, and hair removal. The amplitude and wavelength of the laser's light determines its ability to cut tissue, or stones – lasers are often used these days to cut diamonds. As with all EMR, shorter wavelengths with higher frequencies produce laser beams with higher power. At a certain intensity, a laser light beam will have enough energy to break chemical bonds and melt or even completely vaporize an object, making them the ideal weapons for science fiction. Bottom line: Focused EMR can be extremely powerful.

There are many physiological processes that occur when living organisms are exposed to nonionizing EMR. In this chapter, we discussed the visible light spectrum, and a small portion of the ultraviolet spectrum. From photosynthesis and vision to vitamin D production, light has enabled life on Earth and shaped the physiology of all living organisms. It is the repetitive cycling of light and dark that entrains us to the Earth and all other living beings. But our lifestyles have disrupted our natural circadian rhythms and those of other living beings exposed to our artificial lights at night. The combination of decreased melatonin production and diminished vitamin D production have affected our bodies' systems, including the immune system, and increased our risk of disease. But simple lifestyle modifications can help resolve this imbalance. In the next chapter, we will take a look at how lower frequencies of human-generated radiation can also affect the physiology of cells and human organ systems.

CHAPTER 2

E = MC^2: THE RELATIONSHIP OF ELECTROMAGNETIC ENERGY TO OUR BIOLOGY

The Electromagnetic Spectrum

In one of his most famous experiments from the 1660s, Sir Isaac Newton passed a beam of light through a prism and proved that all of the colors of the rainbow were contained within white light. Although it wasn't clearly understood at that time, the separation of white light into its component colors was proof that each color has its own wavelength and frequency. This was implied by this simple experiment because when waves refract by passing from one medium into another, energy with longer wavelengths will bend to a lesser extent than shorter wavelengths. It is for this reason that the shorter, blue wavelengths project along the lower part of the rainbow while the reds are located above.

It was therefore recognized that visible white light is actually a *spectrum* of varying wavelengths, but the boundaries of this spectrum were not yet defined. While working with prisms and sunlight, the astronomer Sir Frederick William Herschel discovered in 1800 that the temperatures of different colors of light increase as one goes from the violet end to the red end of the visible light spectrum. Importantly, Herschel also noticed that *beyond* the red-light band, there was an even higher temperature. In making this observation, he unintentionally stumbled on the existence of what we now call *infrared energy*. His discovery was quickly followed by experiments performed by Johann Ritter in 1801, in which *ultraviolet* light wavelengths were identified beyond the violet band of the visible light spectrum. These two discoveries were extremely important because they indicated that there was invisible energy coming from the sun that was related to the light that we can see. This generated curiosity among the scientific community about whether there might be even more wavelengths that our senses could not detect.

Although visible light–and the ultraviolet and infrared bands above and below it–had been identified as being composed of waves, the character of those waves was not yet understood. In the 1820s, a scientist named Hans Christian Oersted discovered while working with electricity that electrical currents generate magnetic fields. His research and that of Michael Faraday provided insight that electromagnetism and visible light are related. We now know that light and related forms of energy are composed of two interrelated waves, one with an electrical component and the other with a magnetic component (Figure 2.1). The combination of these two fields, which are oriented at 90 degrees off axis to one another, is referred to as an *electromagnetic field*, often abbreviated by the acronym EMF. Collectively, this form of energy is called *electromagnetic radiation*, or *EMR* for short.

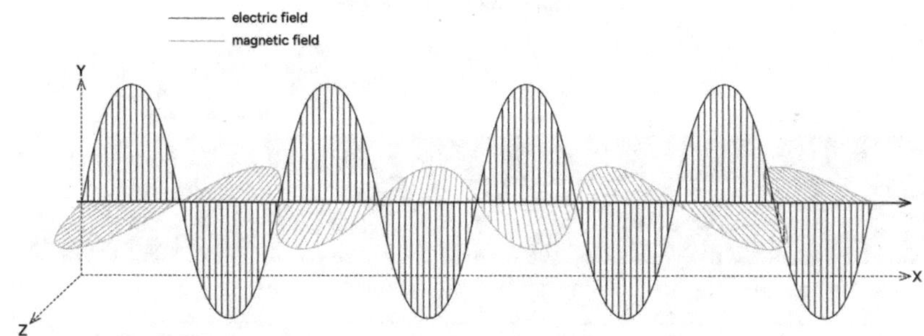

Figure 2.1. Waves of electromagnetic radiation

In 1864, James Maxwell predicted the existence of yet another invisible form of electromagnetic energy wave called *microwaves*, and soon after proposed that visible light waves were a form of EMR as well. It took over twenty years, but in 1888, the German physicist Heinrich Hertz proved that Maxwell's theory was correct and microwaves do exist. The band of radiation called *X-rays* was discovered by Wilhelm Roentgen in 1895. And, in 1914, Earnest Rutherford's discovery that gamma rays, too, were a form of EMR put the final and highest-energy segment of the electromagnetic spectrum into place.

Today, we understand that all of these forms of EMR are related to one another. The differences between them are their *frequency* and the frequency's corresponding *wavelength*. This continuum of energy, referred to as the *electromagnetic spectrum*, extends from radio waves, whose wavelengths range from 1 millimeter up to 100,000 meters (sixty-two miles) in length to gamma rays with ultrashort wavelengths of 100 picometers, or 1.0×10^{-10} m (a picometer is one trillionth of a meter). This enormous span is divided up into categories, typically referred to as radio waves, microwaves, infrared waves, the visible light spectrum, ultraviolet rays, X-rays, and gamma rays (Figure 2.2).

The Electromagnetic Spectrum

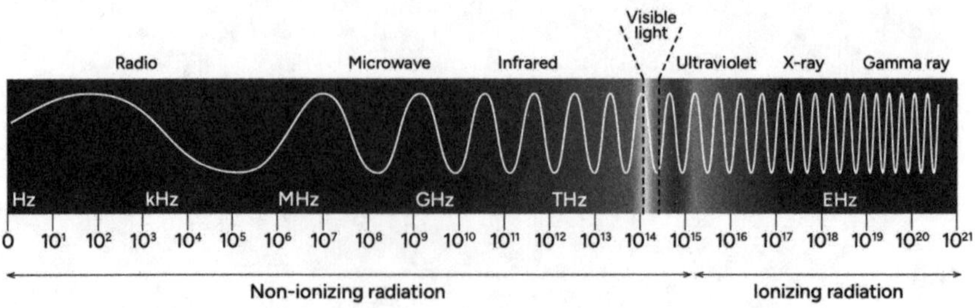

Figure 2.2. The electromagnetic spectrum

Because the speed of "light," which applies to all electromagnetic waves, is constant within a given medium, there is a reciprocal relationship between frequency and wavelength. Multiplying one by the other will always equal the speed of light, as represented in the following formula:

$$c \text{ (speed of electromagnetic waves (light))} = \nu \text{ (frequency)} \times \lambda \text{ (wavelength)}$$

Frequency and wavelength are therefore inversely proportional to one another. As one goes up, the other goes down. Frequency of an electromagnetic wave can easily be determined if you know the wavelength by using the formula

$$\nu \text{ (frequency)} = c \text{ (speed of light)} / \lambda \text{ (wavelength)}$$

Identifying a specific electromagnetic wave can be done by referring to its wavelength or frequency. By convention, longer waves are usually expressed by their frequency, whereas higher energy waves are referred to by their wavelength. There are exceptions to this rule, however. For example, a Wi-Fi router is often identified by its frequency–a 2.4-gigahertz (GHz) router emits radiation with a frequency of 2.4 billion waves every second–even though typical microwave ovens, which also emit radiation at 2.4 GHz, are referred to by the ultra-short wavelength they use.

If you are unfamiliar with the concept of waves, please refer to appendix A (page 319), which provides a helpful brief introduction to wave theory. Because the range of frequencies in the electromagnetic spectrum is so enormous, multiple prefixes are used to simplify their identification. Frequencies (in cycles per second) are described with the unit hertz, in honor of the scientist by the same name. The most common prefixes used include:

Prefix	Abbreviation	Meaning	Example
Kilo	K	One thousand	kHz
Mega	M	One million	MHz
Giga	G	One billion	GHz
Tera	T	One trillion	THz

EMR is ubiquitous, a universal form of energy. Every point of light in the night sky is either generating or reflecting EMR. Our sun, the closest star to planet Earth, is in effect our solar system's EMR generator, providing us with a nearly unlimited source of energy. We absorb EMR from the sun, while at the same time, we are immersed within the Earth's subtle pulsations, called *Schumann resonance frequencies*, and magnetic field. These energy waves create an invisible grid enveloping the earth within which life has evolved and been designed to navigate over billions of years. Living beings also produce EMR, which can be sensed by other organisms. We were created in an electromagnetic world some call the Garden of Eden.

Technology has enabled humans to produce and harness electromagnetic energy of all different frequencies and wavelengths on demand and with purpose. In this book, we will be mostly concerned with the radio spectrum, which extends from 3 Hz to 300 GHz. This immense expanse of energetic frequencies has been subdivided into additional subcategories, many of which will be referred to throughout this book:

Referred to by frequency:

Extremely low frequencies (ELF-EMF): 1 Hz to 300 Hz
Very low frequencies (VLF): 3 kHz to 30 kHz
Radio frequency (RF): 3 kHz to 300 GHz

Referred to by wavelength:

Microwaves (MW): 300 MHz to 300 GHz
Millimeter waves (MMW): 30 GHz to 300 GHz

It may be hard to grasp that energy with characteristics as different as X-rays and radio waves are related in that they are both forms of EMR. It is only the frequency and wavelength of EMR that gives waves their character and determines how they interact with matter, including living beings.

EMR Transmits Energy

One of the reasons different categories of EMR have different effects on matter is because the photons of each frequency carry their own unique amount of energy. The energy of a photon associated with a particular frequency can be calculated by the formula

$$E \text{ (energy)} = h \text{ (Planck's constant)} \times v \text{ (frequency)}$$
$$\text{Planck's constant} = 4.1357 \times 10^{-15} \text{ eVs}$$
$$= 6.626 \times 10^{-34} \text{ Joules}$$

This formula states that the energy transmitted by EMR increases as frequency increases. Therefore, ultraviolet frequencies transmit a greater amount of energy than infrared frequencies. Infrared frequencies transmit more energy than microwaves. In addition to the energy of a single photon, the total energy (or *power density*) transmitted by a wave will be proportional to the square of its *amplitude*. Essentially, the wave's power density is a measure of the number of photons that make up the wave. Thus, a microwave oven will produce more energy than a low-voltage lamp, even though the visible light spectrum is a higher-frequency bandwidth than microwave radiation! This is because the power density of the radiation created by the microwave oven is much greater than that generated by a lamp.

Ionizing Radiation

As described in the previous chapter, at approximately midway through the ultraviolet frequency bandwidth, the energy transmitted by EMR is great enough to potentially eject an electron from its atomic orbit and cause *ionization*. Just like with resonance, ionization is a relationship between energy and matter. Whether or not ionization will occur is dependent not only on the energy of the EMR wave but also on the affinity the impacted electron has for its host atom. Each element has a unique number of protons exerting force on its electrons, and therefore its electrons have their own unique binding affinities. The alkali metals that form salts, including lithium, sodium, and potassium, are among the most easily ionized atoms.

Water molecules can also be ionized. Interestingly, the electron binding affinity for water molecules is not constant but, instead, has been shown to be variable and related to the water molecule's position within the volume of water![26] In living systems, the binding affinity for electrons and the potential for ionization can also be influenced by the presence of nearby metals. While the ionization characteristics of water are uniquely complex and maneuverable, it is only necessary to understand that the ionization parameters for water are not set in stone. It is not necessary for our purposes to know which atoms are at greatest risk for ionization with a specific frequency of radiation.

Physicists have long believed (and many still believe) that aside from generating heat, ionization is the only potential harmful physical effect that EMR can have on living cells. Because ionization is believed to be limited to those higher frequencies from the mid-ultraviolet B (UVB) range through gamma radiation, forms of EMR with lower frequencies have been deemed "safe." This is an extremely important point, because this is the scientific assumption used to green light all of our current technology. We are instructed to cover ourselves daily with sunblock to protect ourselves from ionizing ultraviolet radiation, while at the same time, we are immersed in lower frequencies of EMR that are thought to be of no concern. This may make sense in the theoretical world of physics, but does it make sense when considering the complex biology of living organisms? Let's take a deeper look into the biological effects of visible light, a common source of nonionizing radiation, before exploring the more expansive radio bandwidth.

What happens to electromagnetic energy when it is absorbed by a human body?

This is the big question. I wish I could write this chapter with precision and certainty, as if this were all decided, accepted science, but it is not. At least, not yet. In the previous chapter, we discussed the excitation of electrons that occurs when they absorb photons of light, leading to the creation of sugars during photosynthesis. Numerous biological effects have been observed following the exposure of living organisms to electromagnetic frequencies in the extremely low-frequency

(ELF)-EMF, very-low-frequency (VLF), low-frequency (LF), and radio-frequency (RF) bands. But the lower frequencies of the electromagnetic spectrum themselves create an enormous compilation of energy, and each frequency may potentially cause its own effects. For almost a century, studies have been performed on all forms of living systems, from bacteria and cell cultures to plants, insects, animals, and humans. Experimental data and conclusions have been written up in thousands of peer-reviewed publications. Some effects have been observed consistently while others have not. Over time, though, we are filling in the missing puzzle pieces, and a clearer picture is emerging.

Stochastic Versus Deterministic Effects of Radiation

The clearest picture we have about the health effects of radiation come from the observations of ionizing radiation exposure on living organisms, both in the lab and in the real world. The bioeffects of radiation have been divided up into two categories: *stochastic* and *deterministic*. The distinction between these categories is important to grasp because it explains many of the inconsistencies in the scientific literature and in human experience.

Deterministic effects are perhaps the easiest to understand because they represent the predictable cause and effect with which we are most comfortable. With ionizing radiation, a certain exposure will result in a severe burn. We can apply the same concept to sunbathing. Although there are different tolerances to the sun's radiation based on the concentration of melanin in the skin, for a given skin type, a given length of sun exposure will predictably cause a sunburn. This is a reproducible phenomenon, meaning it will happen over and over again. If a doctor shines a light onto your eye, the iris will constrict, causing the pupil to become smaller. This is a deterministic effect caused by the light.

Stochastic effects are not predictable. They may occur on Monday, but not on Tuesday. They may occur in some people, but not in others. They do occur with increasing frequency as exposure increases. We know stochastic effects are real because we see them with increasing frequency when large populations are studied. If the sample size is large enough, a percentage of risk can be calculated for a specific condition. But it is not possible to determine whether an individual will definitely get or not get a specific disease from exposure. The development of cancer is a perfect example. If someone is exposed to excessive sun for many years, they may or may not eventually develop skin cancer. The reasons are variable and dependent on the person's underlying health and immune status. But while radiation certainly causes cancer, we cannot predict that it will cause cancer at a given point in time in any one individual. Although we try to limit a patient's exposure to ionizing radiation during diagnostic testing with X-rays and gamma rays, there is always a slight chance the person could end up with a disease such as cancer from the exposure.

The mechanisms behind deterministic effects of EMR exposure are fairly well understood. Heating, whether from a cell phone or an X-ray machine, occurs because the energy transmitted by the radiation is absorbed by the body's tissues, increasing its temperature. Simple enough. But the underlying mechanism behind stochastic biological effects ("bioeffects") are not well understood or universally accepted. Although ionizing radiation has been regulated based on the risk for stochastic effects, the same is not true for nonionizing radiation. Inconsistency in experimental results enables the tech industry to maintain their position to the world's governments and regulating agencies that nonionizing radiation does not *definitely* cause *any* biological effect other than heating. Similar to many other industries, promoting uncertainty provides them with a free pass to continue generating more and more EMF-producing devices, utilizing more and more frequencies of radiation, as long as they don't cook us like food in a microwave oven. Have you read *Merchants of Doubt* by Oreskes and Conway? If not, I highly recommend you do. It is a fascinating eye-opening nonfiction book detailing how the tobacco and other industries successfully create controversy in order to manipulate public opinion and steer government regulation to their benefit.

Regardless of the smoke and mirrors, nonionizing radiation has been demonstrated to cause important biochemical effects on cells. Mechanisms of action have been proposed to explain these observations and their impact on cellular health.

Does EMR Cause Cancer?

The most frequent question I hear from people is: "Does cell phone radiation cause cancer?" Before addressing this question, let's start by stating accepted science. As just mentioned, we know that ionizing radiation in the short wavelengths of UV, X-ray, and gamma rays can cause cancer. As such, regulatory agencies worldwide control the production of ionizing radiation and the population's exposure to it.

Through a poorly understood mechanism, it is believed that ultraviolet radiation can cause skin cells such as melanocytes, specialized skin cells that create pigmentation, to go haywire and begin reproducing uncontrollably, resulting in a skin cancer. The tumor produced by cancerous melanocytes is called *malignant melanoma*. UVB radiation theoretically has enough energy to ionize an atom in a skin cell's DNA, change it biochemically, and cause a mutation, potentially leading to cancer. As a society, we believe that the sun, and in particular UVB radiation, is *the* cause for melanoma. We accept that a small bandwidth of radiation coming from the sun and passing through the atmosphere is dangerous and causes a potentially deadly skin disease. We have technology to protect ourselves from this threat, that is, sunscreen.

Yet, there are inconsistencies in this theory. Are you aware that people develop melanoma in areas "where the sun don't shine?" People develop melanoma in the butt-crease, between the toes, and on their retina. One would think the retina in particular should be protected from ultra-short wavelength bands of ultraviolet radiation. After all, light and higher-frequency nonvisible bands of radiation must pass through either the bony structure surrounding the eye or through the cornea, lens, and vitreous fluid of the eye before coming into contact with the retina. These superficial structures should absorb these higher-frequency ultraviolet bands of radiation.

Even though all of the pieces don't fit together, the scientific community, industry, governmental regulators, media, and therefore the rest of us accept that UVB is dangerous and that we should wear sunblock and UVB-blocking sunglasses.

This may not be a big deal to many of you, but, as a melanoma survivor, I am bothered by the inconsistency. Although I could certainly attribute my particular disease to excessive sun exposure, it doesn't explain the increasing incidence of melanoma worldwide, particularly given the industry's ability to get so many people to put higher sun protection factor (SPF) products on their skin and apply those products with increasing frequency. I believe other factors are at work. Perhaps chemicals in skin care products are damaging the skin and causing increased susceptibility to melanoma? Perhaps the false sense of security of wearing sunblock results in people spending more time out in the sun? Even if that were a contributory factor, it still wouldn't explain why melanoma can develop in areas that have not been exposed to the sun. Mysteries remain.

An interesting scientific paper came out in 2000 in which scientists showed a correlation between the incidence of melanoma and the proximity of radiofrequency modulation (FM) broadcasting antennas. By studying the demographics of melanoma in four countries, the authors found that higher numbers of FM transmitters in an area was associated with an increased incidence of melanoma.[27] But how could this be? FM radio broadcasts transmit through *very low frequencies* of radiation in the MHz range, much lower frequencies than ionizing UVB. In fact, FM transmissions are at frequencies lower than visible light, infrared, and microwave radiation. This paper came out over twenty-three years ago, and although the epidemiology described is fascinating, there has been no further published research into this potential correlation.

Maybe the authors didn't take one or more confounding variables into account? That's certainly possible, but what if there really *is* a connection between radio transmitters and the occurrence of melanoma? What would the societal and financial impacts be if radio broadcasting antennas were proven to cause a disease, including a terrible cancer such as melanoma? Would this discovery unravel the telecommunications industry and wreak havoc on military communication? For reasons that probably seem obvious, this would never be allowed to happen. Not only would confirmation of this relationship affect national security, a discovery such as this would also rock the scientific world by introducing a crack in the foundational understanding of biophysics.

In the eyes of the physicist, only ionizing radiation can affect the structure of an atom. Therefore, only ionizing radiation can *potentially* lead to DNA mutations and cancer. Lower EMR frequencies do not harness enough energy to cause ionization. Therefore–using the physicist's paradigm–only higher-frequency UVB rays are harmful and potentially carcinogenic, whereas radio waves are harmless. The division of EMR into ionizing and nonionizing radiation may make sense in a physics lab, but using this demarcation as a minimum threshold for biological effects is simply inaccurate and flies in the face of scientific observations of biological systems.

Observable Effects from Nonionizing Radiation

Some effects have been observed and described in the literature as occurring in some people when they are exposed to nonionizing forms of radiation. Bioeffects that are invisible are easy to dismiss, but those that can be observed are harder to deny. One of the first conditions described occurred in people exposed to early CRT screens who broke out in rashes. This phenomenon was only described in Nordic populations in 1994 and was called *screen dermatitis*.[28] This phenomenon was not explainable at the time. More recently, some scientists, including me, have shown that when some people are exposed to microwave radiation, the physical appearance of their blood changes, with visible clumps forming.

Normally, red blood cells flow within the blood plasma as unique biconcave discs. Both their shape and their separation from one another are important to optimize surface area for oxygen and carbon dioxide exchange. But researchers have observed that red blood cells deform and aggregate into an assembly resembling a stack of coins, called *rouleaux formation*, in some people following exposure to the microwave radiation emitted by a cell phone. Although demonstrating this effect requires careful attention to technique, this appearance has been observed by several investigators during live blood cell analysis utilizing dark-field microscopy as well as on peripheral blood smears.[29, 30, 31] More recently, the development of rouleaux from cellphone exposure was documented using diagnostic ultrasound.[32]

This effect is not observed in all people, but when it occurs, it's visually striking (Figure 2.3). More importantly, this abnormal clumping can affect the blood's function. Aside from decreasing the blood's efficiency of gas exchange, rouleaux formation may increase the likelihood of developing blood clots,[33] a condition that can have profound health complications, including stroke, pulmonary embolism, and possibly death. Rouleaux formation is a reversible phenomenon that can result from many different causes, including microwave exposure.[34] It's not clear how microwave radiation leads to abnormal blood formations, but one possibility is the cell's membrane potential–known as the *zeta potential*–changes when it absorbs the radiation.

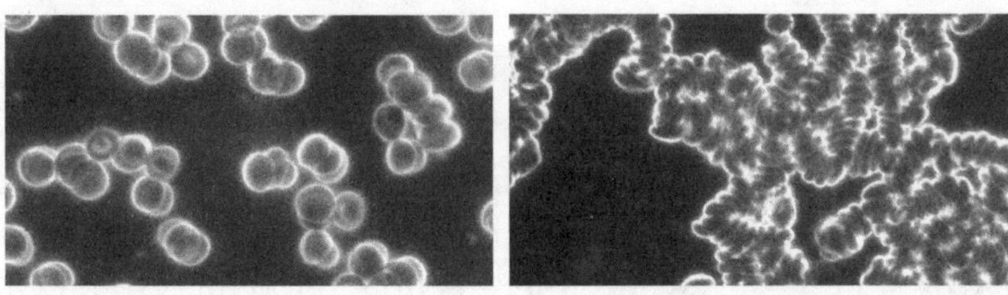

a. Normal blood smear prior to exposure

b. Same person/same day/same hour after 10-minute exposure to 2.45-GHz Wi-Fi router

Figure 2.3. Rouleaux formation. Courtesy of Magda Havas, PhD

If we can *see* an effect–such as red blood cells coming in and out of rouleaux formation–how likely is it that nonionizing radiation from a cellphone or Wi-Fi router has even more biological effects that we *can't* see?

Porphyrins

The mechanisms by which nonionizing radiation causes cellular effects are, in most cases, not decided science. Photosynthesis is an exception. A closer look into this process leads us to focus on the biochemical structure of molecules called *porphyrins*. A particular characteristic of these molecules is essential for photosynthesis to occur, but I believe it may also explain some bioeffects from lower frequencies of nonionizing radiation.

Biologists have long known that EMR in the visible light spectrum can be absorbed by electrons in a chlorophyll molecule. The excited electrons are "bumped up" into higher orbitals and then systematically processed by the chloroplast's machinery in a cascade of events that lead to the production of sugar (Figure 2.4).

$$6CO_2 + 12H_2O \xrightarrow[\text{green plants}]{\text{light}} C_6H_{12}O_6 + 6O_2 + 6H_2O.$$

carbon dioxide   water   glucose oxygen water

Figure 2.4. Photosynthesis

The building blocks of sugar (glucose) are carbon dioxide and water. Naturally, this chemical reaction does not occur anywhere and everywhere there is an interface between carbon dioxide and water. If it did, sugar would rain from the sky on cloudy days! No, the process requires help from other molecules arranged in a specific three-dimensional configuration, as occurs in the chlorophyll molecule. In order for this reaction to occur, electrons need to be loosened up from their parent atoms/molecules and manipulated as currency. This relaxation of the bond between the electron and its host atom is achieved by a precisely and strategically positioned magnesium atom within the center of the chlorophyll molecule. With the bond between atom and electron relaxed, even the low energy contained within a light photon can cause the electron to excite and jump into a higher orbital where they are more easily removed.

Chlorophyll is not the only molecule with a centrally positioned metal ion. Many similar molecules are found in animals.

These metal-containing molecules are called *porphyrin rings* (Figure 2.5). I suspect different metals are used to manipulate the binding affinity of electrons depending on the biological need and desired effect. Red blood cells contain heme, a precursor to the molecule hemoglobin. Heme houses an iron atom. It is the iron atom's effect on the binding of electrons that allows hemoglobin to bind and release oxygen and carbon dioxide in our red blood cells. Without metal-containing porphyrins, I don't think life could exist.

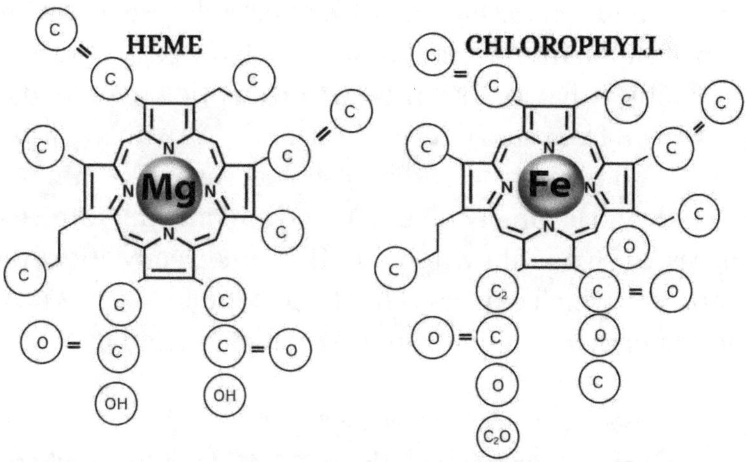

Figure 2.5. Porphyrin rings

There are many different porphyrins in the human body, each with crucial roles. It is therefore important to fortify our diet with trace minerals to ensure proper porphyrin production. It seems plausible that one of the ways that low-frequency radiation may impact biology is by "relaxing" electron bonds near metal ions, although this has not yet been determined.

Cells

Plants and animals are created by the coordinated communication of billions and even trillions of tiny microscopic subunits called *cells*. Each cell has components called *organelles* that allow the cell to function and carry out its business. In addition, with the exception of red blood cells, each cell has its own central control center called the *nucleus*, which houses the DNA.

Each cell is surrounded by a membrane that keeps it intact. An important function of the cell membrane is to separate a cell's internal environment from the external world, much like our skin protects our body. By controlling the cell's interior, a predictable and hospitable environment is created, allowing the cell to carry out its functions. A functioning cell membrane is required for life. In 2007, an Israeli researcher named Joseph Friedman showed that when cells are exposed to EMFs, they exhibit behavior as if they are attempting to defend themselves, even from short-term exposure![35] As with an organism, if a cell is constantly defending itself, its resources can become depleted, and it can become stressed and malfunction. In addition, research has shown that EMFs can directly affect the cell membrane, which can also lead to cellular stress. It is an interesting phenomenon that requires some detailed explanation.

Stress

Stress is a word we hear all the time, and yet few truly understand its meaning and implications. At its most basic definition, stress occurs when a repetitive force is applied to a living or inert "thing" that over time compromises its integrity, structure, and organization. Stress is different from *trauma*, which involves a force with enough power to break something apart on its first encounter.

As an example, consider that a bone can break from a traumatic injury, such as falling from a height or suffering an accident while traveling at a high velocity. A bone breaks from trauma when the force applied to the bone exceeds the binding strength of the cells and matrix holding the bone together. Not everyone will suffer a fracture from the same injury because our bones have different strengths and elasticity. Lesser, seemingly inconsequential, injuries can also affect the integrity of a bone. Consider someone who has been relatively sedentary and starts a new jogging routine. If the jogger doesn't take breaks to allow the bone to heal from the *microtrauma* that occurs during each day's run, the bone can weaken and, over time, fracture. The force of the injury occurring each day isn't great enough to cause a traumatic fracture, but the gradual accumulation of injury over time leads to the same result, a broken bone–in this case called a *stress fracture*.

This is an abstract concept that can be applied to other arenas. Stress can affect the enormity of a galaxy as well as the minuteness of a cell because the structures of both large and tiny objects are held together by organized energetic forces that can be weakened. In correlating this concept to the world of EMR, consider ionizing radiation to have the energy to cause a traumatic injury, ionization, whereas nonionizing radiation has the capability to cause stress-related injury.

But how can a cell be under stress?

Mechanism for Cellular Stress

In order for a cell to be whole and healthy, it needs to have a properly functioning membrane. Part of the cell membrane's job is to control the entry of materials into the cell and to eliminate waste no longer needed by the cell. Cell membranes contain pores or channels that selectively open and close, choosing when and if specific ions and molecules can gain entry. Ion channels open and close in response to the cell's need to keep its internal environment at proper concentration levels. The concentration of intracellular calcium, for example, is regulated by ion channels. By creating a concentration gradient of ions like calcium between the exterior and the interior of the cell, a relatively strong voltage potential is generated across the cell membrane. One of the major differences between a cell that is alive and one that is dead is the ion concentration gradient across its cell membrane.

Increased Intracellular Calcium

In the early 1990s, biologists discovered that ELF-EMF can result in an increased concentration of calcium ions in the cell and its organelles, including the mitochondria.[36] Since then, both ELF-EMF and RF-EMF have been shown to cause increased levels of intracellular calcium. This discovery was curious and important. Although the concentration of intracellular calcium could be measured in a laboratory, the mechanism *causing* this state was not obvious. When a study performed by the biochemist Martin Pall, PhD, showed that pharmaceutical-grade calcium channel blockers could prevent this effect, it seemed that somehow ELF-EMF and RF-EMF were opening calcium channels on the cell membrane and allowing calcium to flow into the cell.[37] This was an important discovery because it illuminates one way–though there are probably others–nonionizing radiation can affect an individual cell's biochemistry. Also importantly, the phenomenon is reproducible.

Increased intracellular calcium can lead to significant biological repercussions. Since the initial discovery that EMFs can activate voltage-gated *calcium* channels, we have learned that EMFs can also open voltage-dependent *anion* channels, which may explain red blood cells' tendency toward rouleaux formation.[38] Membrane depolarization may be the common mechanism causing both phenomena.

The impact of increased intracellular calcium on a cell's health is dependent on several factors, which lead to different outcomes. According to the theory by Dr. Pall, increased intracellular calcium can stimulate biochemical pathways that lead to the production of nitric oxide. Nitric oxide can have a *beneficial* effect, which may explain some of the therapeutic responses observed with EMF therapy. But nitric oxide can also be a precursor to a molecule called *peroxynitrite* and lead to an excessive production of free radicals in the cell, a condition called *oxidative stress* (Figure 2.6).

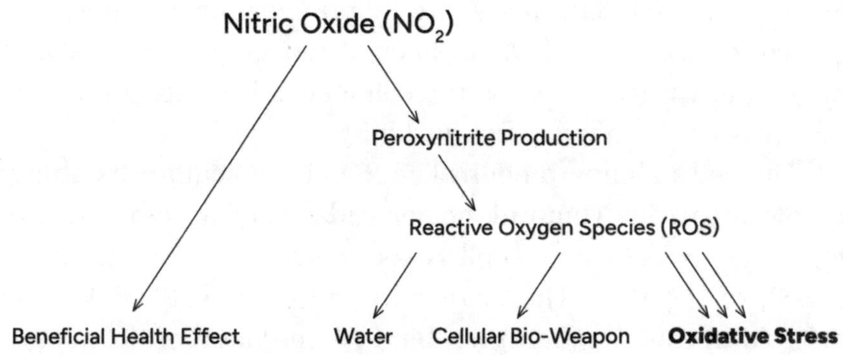

Figure 2.6. Nitric oxide pathways

These two pathways exist in our cell's machinery because both end products are potentially useful. Nitric oxide can have beneficial health effects, such as causing the vasodilation of small blood vessels. The generation of free radicals, called *reactive oxygen species* (ROS), is also an extremely important part of normal cellular function. However, the accumulation and distribution of ROS need to be under the cell's control. If the number of ROS gets out of hand, the cell becomes stressed.

There are three main ROS–superoxide ($*O_2^-$), hydrogen peroxide (H_2O_2), and hydroxyl radical ($-OH$). These are unstable molecules. Mitochondria, the powerhouses of the cell, commonly produce ROS, but these organelles are supplied with enzymes that break down these dangerous ions into nonreactive compounds. The enzyme superoxide dismutase 2 (SOD2) converts $*O_2^-$ into hydrogen peroxide (H_2O_2). Catalase and glutathione peroxidase are two other enzymes that can fully reduce and convert H_2O_2 into water and oxygen. If H_2O_2 is incompletely reduced, -OH will be created, effectively producing one ROS from another. Each ROS is potentially dangerous and reactive and can cause damage to the cell.[39]

The formation of ROS is a normal part of our physiology. Some cells, including the white blood cells, neutrophils, and macrophages, store ROS in specialized compartments and may rapidly release them if the cell feels it is under attack.[40, 41] This process is called *oxidative burst*.

The cell's ability to neutralize ROS is critical for its ability to tolerate toxins. Glutathione peroxidase facilitates the transfer of electrons from the cell's master antioxidant, glutathione (GHS), to the ROS. The oxidized glutathione (GS) molecules created in this reaction then bind to one another, forming a disulfide bond (a bond between the two sulfur atoms). The resulting molecule is abbreviated "GSSG" (Figure 2.7).

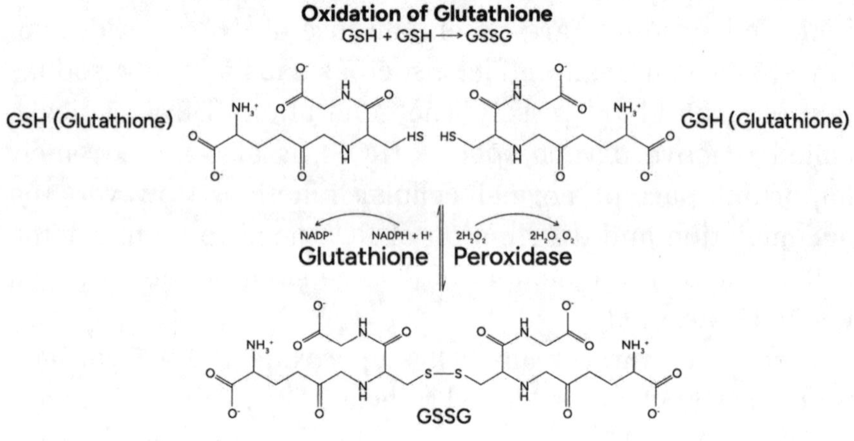

Figure 2.7. The biochemical reaction of glutathione

Antioxidants, such as vitamin C, generously donate their electrons to molecules like GSSG, allowing them to dissociate back into two GSH molecules. If there are insufficient electrons available from other antioxidants to reduce GSSG, GSH levels in the cell drop and ROS build up. The result, oxidative stress, can be very damaging to cells. Measuring glutathione levels is one way to determine the extent of oxidative stress in a tissue sample.

If ROS were only formed by normal cellular processes, we would be in good shape. But unfortunately, many environmental toxins have been shown to generate ROS and raise cellular levels. Among this extensive list are tobacco,[42] pesticides,[43] some drugs,[44] and heavy metals, such as nickel, arsenic,[45] mercury,[46] and cadmium.[47] In addition, both ionizing radiation[48] and nonionizing EMF[49] generate ROS.

Repercussions of Oxidative Stress

At this point, one might be asking, *How can oxidative stress damage a cell?* and *How can a cell break?* The answer lies in biochemistry. ROS have a strong need to be reduced and acquire electrons to form stable atoms. If there is no glutathione available, ROS may look to steal electrons from other molecules within the cell, including the fat molecules–also referred to as lipids–that compose the cell membrane, nucleic acids like DNA, and any cellular proteins. When this occurs, the ROS stabilizes, but a new oxidized, positively charged ion is formed from the molecule that was robbed. This new ion will be unstable and will look to grab an electron from another atom or molecule, generating another chemical reaction, and the process repeats. This cascade of events can cause significant cellular damage, depending on which structures are involved.

Mitochondrial Dysfunction

In order for cells to accomplish tasks and keep themselves intact and functioning properly, they need a source of energy. Adenosine triphosphate (ATP) is the primary molecule cells use to store and utilize energy. Think of ATP as a form of energy currency. When enzymes break this molecule apart, the reaction generates predictable bursts of energy that the cell's components can then use to get things done. The production of ATP occurs in the *mitochondria*.

If ROS take electrons from molecules within a mitochondrion, they can damage the organelle with significant repercussions for the entire cell. Mitochondrial dysfunction can reduce ATP production, the cell's biochemical currency, and debilitate the cell by decreasing its energy supply. This is a big deal. Weak cells can lead to weak organs. Mitochondrial dysfunction is an underlying factor in many diseases, some of which will be discussed in future chapters.

Lipid Peroxidation

ROS can also steal electrons from fatty acids in the cell membrane. This effect may not be too foreign to you. The same chemical reaction, oxidation, causes edible fats like seed oils to go rancid. ROS such as hydroxyl radicals and peroxynitrite molecules can cause oxidation of fatty acids within the cell's membrane. When fat molecules lose an electron, they change structure and become an unstable molecule called a *hydroperoxide*. The peroxidation of a membrane lipid can lead to a cascade of events in which one lipid steals an electron from its neighbor, and so on, causing peroxidation of many lipids in the cell membrane. The end products of this degradation are *reactive aldehydes*. This process can damage the integrity of the cell membrane, alter its permeability, and, in some cases, impact the cell's function and viability. Some labs measure aldehyde levels to assess the severity of oxidative stress.

Protein Damage

Proteins can also be targeted by electron-seeking ROS when a cell is in oxidative stress. Proteins, big three-dimensional macromolecules composed of a chain of amino acids, are important for both cellular structure and function. ROS can steal electrons from amino acids, the building blocks of proteins, and therefore cause protein oxidation. There are twenty different amino acids, but the two most commonly oxidized by ROS are the sulfur-containing cysteine and methionine.

A protein's shape is crucial to its function. When an integral amino acid is oxidized, the protein's shape can change, compromising its function. This is particularly true for enzymes, a special class of proteins that catalyze chemical reactions. The term for this process is *denaturation*. Denaturing can render enzymes useless. In addition to oxidative stress, tissue heating can cause protein denaturation. When denatured, an oxidized protein can become fragmented or bind to another protein through the creation of protein-protein cross-linkages.[50] As will be discussed, this is how Lewy bodies are formed in some neurodegenerative diseases like dementia and Parkinson's. In general, not only can denaturation change a protein's ability to function properly, it can also impact the survivability of the entire cell. En masse, it can even affect an organ's function and cause systemic disease.

DNA Damage

Strands of genetic code are encrypted within DNA located in the cell's nucleus. Genes are composed of sequences of four specific protein subunits called *nucleotides*. Each DNA molecule contains two strands of nucleotides that loosely adhere to each other. In order for this to happen, complementary nucleotides on each strand bond to one another. Each pair of bonded nucleotides is called a *base pair*. The conformational shape of DNA when base pairs are bonded to one another resembles a spiral staircase and is termed a *double helix*.

When ROS enter the nucleus, they can steal electrons from nucleotides and damage DNA. The damage caused by oxidation is variable and dependent on the gene involved. In fact, extensive research into the effects of various "base lesions" is ongoing, but this is beyond the scope of this book. Nuclei have developed multiple overlapping mechanisms to repair damaged DNA. Damaged nucleotides can even be removed and replaced. ROS can cause epigenetic damage,[51] potentially affecting genetic expression, which can be passed down to offspring. Single-strand and double-strand breaks have also been shown to occur from oxidative damage.[52] Double-strand breaks, in particular, can lead to mutations that cause cancer or cell death and have been shown to occur even at low levels of ROS.[53]

One of the repair mechanisms cells developed in response to low levels of oxidative stress is the creation of a protein called *p53*, which acts as a tumor suppressor and reduces the chance of cell damage. The TP53 gene contains the genetic code for p53. In half of human cancers, the p53 gene is mutated or deleted.[54] Denaturation of p53, changing its shape, is an early marker of oxidative stress in Alzheimer's disease.[55]

In Summary

Oxidative stress can take time to become clinically apparent. Like other forms of stress, cellular oxidative stress can simmer, cause dysfunction, and become the root cause of many diseases down the road. Most chronic diseases–including cancer, diabetes, cardiovascular disease, gastrointestinal disease, neurological disease, and lung disease–are the result of chronic inflammation generated by oxidative stress.[56] Because people in developed nations more commonly suffer from these diseases, they have been referred to collectively as *diseases of civilization.*[57]

Many studies have reported oxidative stress occurring in humans when they are exposed to nonionizing radiation. Occupational exposure to radar, including marine radar, decreases glutathione in the blood[58] and increases malondialdehyde levels.[59] An important review Yakymenko et al. performed and published in 2016 evaluated one hundred scientific research papers that studied the effect of low-intensity radio frequencies (RF) and found that 93 percent confirmed that exposure to nonionizing radiation can cause cellular oxidative stress. The authors of this exhaustive study concluded definitively that oxidation caused by low-intensity RF has a high potential for causing disease.[60]

Now that we have a mechanism to explain how nonionizing EMR can affect cellular function, let's take a look at how we are exposed to different forms of this environmental toxin.

CHAPTER 3

RADIATION GENERATED BY ELECTRICITY AND "DIRTY ELECTRICITY"

Easy access to electricity has enabled technology to advance at a dramatic pace and shaped and dominated modern society. In one sense, distribution of electricity has undeniably altered the biology of humans and animals. Electricity and artificial sources of illumination, in particular, have drastically affected our circadian rhythms and knocked us out of sync with Earth's orbit and seasonal variation in daylight. But one could opine that electricity has allowed a significant step in the evolution of humankind and liberated us from the biological constraints created by our natural bonds to Earth. As a physician, my main concern is the physiological health of the human body, so that is where my attention often drifts.

Take a minute to consider that lightbulbs, smartphones, and all the other electronic gadgets you can think of are powered by electrical current, whether it comes from a power grid or a battery. Although we have created a society centered on the applications of electricity, most of us have no idea how it works or how it can affect our health and well-being.

Electricity

Electricity is generated when negatively charged electrons flow through a conductive material. The vast majority of that current flows on the conductor's surface, as electrons repel one another, and seek to spread out as much as they can. Because metals allow free movement of electrons throughout their volume, metallic wires create a suitable conduit for the flow of electrons and therefore the transmission of electrical current. Electrons will only flow down a wire when there is a relative difference in electrical potential between the two ends of the wire. In tribute to the Italian physicist Alessandro Volta, who invented the first electric battery, the unit for quantifying that potential difference from one end of a wire to the other is the volt (V). The word used to describe the difference in potential is *voltage*. The higher the voltage, the greater the difference in charge from one end of the wire to the other, and the higher the power of the electricity. A clear analogy to electrical potential is hydrostatic pressure in pipes that conduct water. More pressure creates more water flow just as more voltage creates more flow of electrons.

Electrical current, like voltage, is an important determinant, and is a measure of the number of electrons that travel through a given cross-sectional plane of the conduit in one second. Current is represented by the letter "I" in formulas and is measured in amperes (A), amps for short, in commemoration of the French physicist André-Marie Ampère, a founder of electromagnetism. Quantitatively, 1 A is defined as 6.25×10^{18} electrons per second. In terms of the water analogy, amperes of current are analogous to units such as gallons per minute of water flow.

The movement of charged particles such as electrons down a wire produces a radiant energy field. The Danish physicist Hans Christian Oersted was the first scientist to discover this magnetic energy field that surrounded the wire transmitting current. James Maxwell further discovered that the energy field around the wire was static and consisted of both electrical and magnetic components. Maxwell also discovered that the magnetic field force had an orientation that was dependent on the current's direction. Maxwell's right-hand thumb rule models this relationship. With fingers of the right hand flexed around a wire and thumb extended toward the direction of current, the fingers point in the direction of the magnetic field (Figure 3.1).

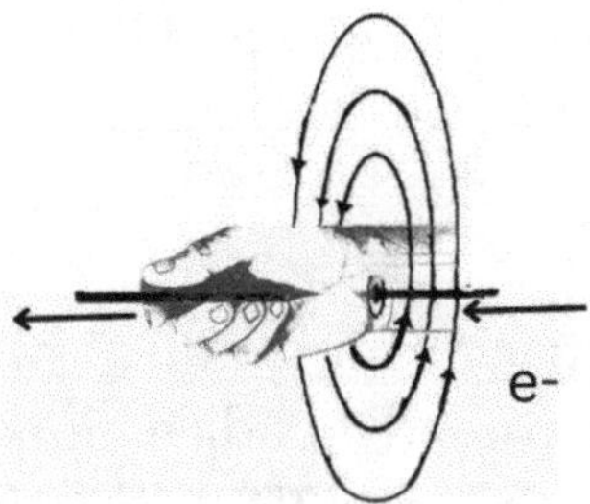

Figure 3.1. Maxwell's right-hand rule: Electrons flowing through a wire in the direction of the pointing right thumb generate a magnetic field in the direction of the fingers.

If electrons encounter obstacles along their path, such as atomic nuclei, they will collide, resulting in a resistance (R) to their flow. With resistance, electrons move at a slower and more erratic pace. It follows then that as the resistance goes up, current will go down. Resistance is measured in units called ohms to honor the German physicist Georg Ohm, who discovered the direct proportionality between voltage and current. The formula that expresses this relationship is called ***Ohm's law***:

$$V = IR \text{ (voltage = current x resistance)}$$

Although we don't want to get bogged down with math and formulas, this is one that is important to grasp. Thankfully, it is a simple concept to understand because it is intuitive.

Technology

The first major invention utilizing electricity was the telegraph, developed in the 1830s and 1840s. By transmitting pulses of electrical current from a sending point, along a wire, to a receiving point, and using those pulses to activate an electromagnet at the receiving end, transmission of information through wires over long distances from one end of the wire to the other was possible. A code of shorter and longer pulses–called dots and dashes–developed by Samuel Morse, was the most common method used to represent numbers and letters of the alphabet. Telegraphy enabled communication between cities and countries but was not a household technology.

Circuits

In the infancy of electricity, Alessandro Volta created an apparatus that produced a steady unidirectional flow of electrons, referred to as a *direct current* (DC). This energy stream could then be used to activate a device. Later, the creation of the switch allowed current to be turned on and off (Figure 3.2).

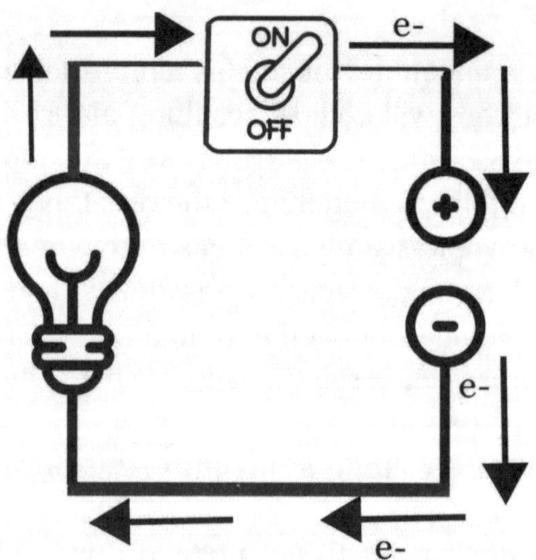

Figure 3.2. Basic electrical circuit

After the invention of the electric light bulb by Thomas Edison in 1879 and the subsequent creation of an electricity distribution system in the 1880s, electrical wiring was brought into the home.

Alternating Current (AC) vs. Direct Current (DC)

While Edison and the General Electric Company were promoting DC power, Nikola Tesla discovered that by repetitively alternating the flow of current–alternately pushing electrons forward through a wire and then pulling them back–waves of energy could be transmitted down the wire without any net movement of electrons. Unlike DC power, this type of electricity produces an *alternating current* (AC) that reverses direction many times a second. The number of times the current oscillates back and forth per second is the current's frequency and is measured in hertz (Hz).

Before the widespread electrification of communities, engineers debated as to which system, AC versus DC, would be a better choice for widescale deployment. DC electricity was preferred by Edison and General Electric and was the original choice for electrical systems in the US. DC current at the time would have come from batteries. But Tesla, with the support of George Westinghouse, implemented AC power at the World's Fair in Chicago in 1893 at a much lower cost than had been proposed by GE. AC power at the time would have come from rotating machines such as electrical generators. Tesla demonstrated those advantages when he successfully electrified Buffalo, New York, which became a model, setting the stage for today's electrical grid on national and global scales. For most countries in the Americas, with the exception of Chile and Argentina, the current in the electrical grids reverses direction sixty times each second (60 Hz). In the rest of the world, with a few exceptions, the frequency of AC power is 50 Hz.

AC power has several advantages over a DC system. Importantly, in an AC system, voltage can be easily "stepped down" or "stepped up" by means of simple, inexpensive transformers. The ability to step down voltage provides the power company with the ability to transmit high voltages, making the production and delivery of electricity much more practical and less expensive than a DC system. In an AC system, high-voltage electricity can be generated at a power plant many miles away from a consumer and be sequentially stepped down to supply a stable lower voltage to individual households and the entire community.

AC electricity is practical and cost-effective, but there is a difference between the static electromagnetic field created by DC current described by Maxwell and the EMFs generated by AC current. Because the direction of current is constantly changing in an AC circuit, the orientation of the electric and magnetic fields also vacillates, generating *waves* of energy instead of a static energy field. The frequency of the electromagnetic waves is the same as the AC frequency, 50 or 60 Hz. This frequency corresponds to ELF-EMF within the electromagnetic spectrum. More on this later.

Power Lines

An AC generation and distribution system allows the utility company to deliver large amounts of power to communities without needing unreasonably thick cables. Because it is the current and not the voltage that is affected by resistance, transformers allow easy tradeoff between low voltage at high current for generation sites and consumer usage and high voltage (but low currents) for long-distance transmission. Step-up transformers therefore increase the voltage from the power plant so transmission power lines can then carry high-voltage electricity across long distances, canvassing the country and much of the world. Transmission lines may be three hundred miles long and carry the highest voltage in the power grid, from 155,000 to 765,000 V, depending on community need. Step-down transformers lower the voltage to supply shorter-range transmission lines that carry voltages less than 100,000 V into neighborhood distribution lines, which themselves typically transport between 2,000 and 34,500 V. By using distribution transformers (Figure 3.3), the electrical power can be further stepped down to supply industrial and residential lines carrying 240 V (two 120-V terminals).

Figure 3.3. Distribution transformer: residential lines and step-down transformers

The possibility that EMFs generated from high-tension power lines could cause negative health and ecological effects that go beyond a low-frequency buzzing noise and strange sensations sometimes experienced when near a power line has been a concern for decades. But, when scientists refer to *high-tension lines*, what category of electrical line are they specifying? In electrical engineering, the definition of a high-tension line is broad and refers to a wire that carries over 1,000 V between conductors and 600 V between conductors and the ground. According to this definition, *all* transmission lines and residential distribution lines carrying electricity are technically high-tension power lines.

Household Wiring

Household wiring has significantly evolved over the years. If you've ever lived in or purchased an older home, you may have been confronted with an electrical system you had not seen before. When I purchased my new (old) home, which was well over one hundred years old, it still had "knob and tube" wiring. This configuration was common between the 1880s and the 1940s and consisted of running two wires–one "hot," one neutral–throughout the home to provide electricity. It was called *knob and tube* because the wires were wrapped around ceramic knobs nailed into wall studs and floor joists to hold them in place and keep them apart from one another. Tubes were placed through wood beams, floor joists, and wall studs to keep the wires from coming into contact with one another and with the abrasive surfaces of wood framing. Knob and tube wiring poses multiple dangers of fire and electrical shock, particularly if the electrical system is overloaded with modern appliances and surrounding insulation has deteriorated. I hadn't personally known anyone who had experienced a fire caused by this old wiring system. But for me, living in an old home fitted with a knob and tube system left me with a subtle anxiety each time I closed my eyes in bed at night.

There have been many improvements and changes in the National Electrical Code (NEC) for electrical systems since that time, importantly including the requirement for a *third* lead that forms a grounded plug. The goal was to provide a safe system, limiting the risk of fire and electric shock. Because of its exposed nature and absence of a ground, knob and tube wiring must be updated in order for a house to pass electrical inspection today.

Nowadays, residential buildings are fed three wires from the utility company: two hot leads and a neutral lead. Most wires throughout a residential facility are tapped to run a voltage of 110/120 V between one of the hot leads and the neutral lead. Outlets can also be wired to provide 220/240 V for large appliances requiring extra power, such as central heat and air-conditioning, clothes dryers, and electric stoves. All 240-V household lines are tapped between the two hot leads. In the breaker box, all hot leads going to 110/120-V outlets throughout the house are drawn from one of the two hot leads at the entrance. All of the neutral leads throughout the house are tied together in the breaker box at a neutral point (Figure 3.4).

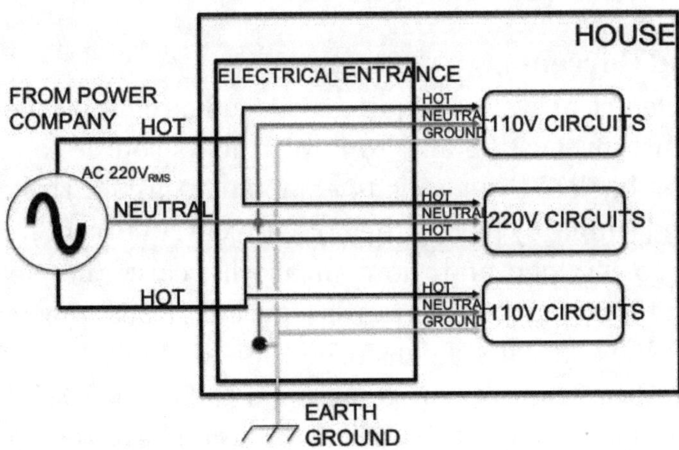

Figure 3.4. Home circuit breaker box: Schematic showing the creation of 110V and 220V circuits in the home breaker box

Electrical wires are bundled within insulated cables, which are typically distributed within the walls, ceilings, and floors of a home. Cables contain multiple wires, each with a different color to allow for easy identification from one end of the cable to the other. By convention, red, black, and sometimes other colored wires are coded to deliver power from the service panel to the outlet or device. White wires are neutral and carry power back to the service panel. Ground wires are either bare, meaning they do not have any plastic coating, or they are colored green. It is important that electrical inspectors ensure that wiring has been connected according to code.

Household wiring has become much more sophisticated since the knob and tube days. An extensive code of standards specifies details such as how far to space outlets around the perimeter of the room and where to place light switches. The NEC is based on safety and is intended to prevent fire, electric shock, tripping hazards from extension cords, overloading of electrical outlets, etc. But unfortunately, the NEC does not consider the health effects of extremely low-frequency electromagnetic fields (ELF-EMF), which have been extensively documented in the scientific literature.

Ground Currents

Every electrical system requires a closed circuit for energy to flow. Current must be able to flow from its source to its load and then back to the source to complete a circuit. This can be achieved simply by having the hot wire carry current from the source to the load and the neutral wire carry the electrons from the load back to the power source. This is how modern 110/120-V electrical systems are designed. At each outlet, the hot lead (the narrow slot) is intended to have 110/120 V on it and the neutral lead (the wider slot) should have zero volts on it. This is implemented at the circuit breaker box, where the power line enters the building.

The Earth, being somewhat electrically conductive, acts as a reservoir for a virtually endless supply of electrons. The voltage at the Earth ground is defined as precisely zero volts. In order to obtain that voltage as a reference, residential installations have a copper rod inserted into the ground near the house, and a wire running from there to the breaker box, where they connect to all of the ground leads of the outlets within the home.

As power consumption has grown, current through household wiring combines with resistance of that wiring, generating electrical voltage drops. Because of this phenomenon, voltage appears on the neutral lead, and is therefore no longer at zero. With neutral leads no longer the same potential as the ground lead, electricity will travel through the ground to try and bring those neutral and ground voltages together. This leads to ground current, a stray voltage, that has become common in both densely and sparsely occupied zones of the grid. Proper wiring precludes connecting the neutral and ground wires together.

Ground current was first noticed in the 1950s on some farms as a contributor to poor health in livestock, particularly dairy cattle. Research indicated that stray voltage caused by the utility company having connected a neutral wire to ground on the farm resulted in a decline in the animals' health status, as indicated by abnormal blood chemistry, and decreased water consumption.[61] Hillman has written extensively about the effects of stray voltage on dairy cattle and has described symptoms including animal agitation, swollen joints, open sores on the animals' legs, foot lifting/dancing, reduced milk yield, low fertility rates, and increased rate of spontaneous death in exposed cattle.[62]

Stray voltage and ground currents can also affect humans. We tend to think that walking barefoot on the earth will be beneficial for us by putting us in touch with the Earth's natural signals. If there is a remnant of stray voltage traveling along the ground between your two feet as you stand or walk, electrical current will pass from one leg, through your body, and down the other leg. This is an illustration of Kirchhoff's law, a law of physics that states that current will travel in any and all paths available to it (not just the path of least resistance). So even if current is happily flowing through the ground, a proportion of it will detour through an animal's body if there is a difference in voltage potential from one foot part of the organism to the other.

At my previous residence, I may have witnessed unexpected effects of ground current. A gas transmission line company owned a right of way down the center of the lot and had installed a large subterranean pipeline. I learned that low-voltage electric currents are commonly applied to gas transmission pipelines to prevent corrosion. Even though the pipes were six feet below grade, the effects of the current, including a heating effect, were witnessed on the surface of the ground. Early during each snow, the path of the pipelines could be clearly seen as the snow above the pipeline would initially melt, creating a tract of persistent green in the otherwise white landscape. Of course, as the snow continued to fall, it would cover the pipeline tract too. The pipeline tract would also be the first area in which the grass would again poke through as the snow melted.

The ground current from the pipeline may also have functioned as an electromagnetic barrier to the invasion of groundhogs. Burrowers had wrought havoc on one side of the property with respect to the pipeline, but in the seven years that I lived there, the critters made no attempt to burrow into the ground on the other side of the pipeline. I suspect it might have been because of ground current. Strong ground currents have also been observed around wind turbine installations in cases where proper safety precautions were not taken.[63]

Electromagnetic Fields

AC power generates ELF-EMF. The electric and magnetic field components are interrelated by sharing the same frequency, but need to be considered independently because their amplitude (field strength) and wave patterns vary from one another. The electric field strength generated by AC power is determined by the voltage of the circuit. As voltage increases, the electrical field intensifies. It therefore makes sense that electric fields produced by long-distance transmission lines are much stronger than the electric fields emanating from neighborhood distribution lines, which in turn are much stronger than the ELF-EMF produced by household wiring. Electric fields from pure AC power are constant because the voltage being transmitted through a given wire is constant.

Current determines magnetic field strength. As current increases, magnetic field strength increases. With increased demand for electricity in the winter or summer months, magnetic fields emanating from transmission and distribution lines will be stronger.[64] The distance the electric and magnetic fields irradiate are dependent on their intensity and whether or not they are absorbed in transit. In general, radiation levels drop by the square of the distance (d) from the source of exposure. Therefore, if you double your distance from a source of electromagnetic radiation, your exposure will be one-fourth of what it had been previously.

Inverse Square Law:
New Intensity = Original Intensity $\times 1/d^2$

Within the home, electric fields are emitted by all unshielded live wires. With field strength falling off with the amount of insulation around them and the square of the distance from the source, those fields dissipate quickly. Electric fields are generally detectable for a distance of one to two feet from the wall housing the wire. When lamps, extension cords, and appliances are plugged in, electric fields emanate from their power cords and, in some cases, the appliance itself, regardless of whether or not the device is turned on. If wires are traveling beneath the floor to a floor socket or an overhead light in the room below, electric fields will travel up from the floor between one and two feet. Not all cabled wires emit electric fields. Insulated wires with shields, such as those encased within mineral-insulated copper-clad cable and metal clad "MC" cable, block electric fields, but such wiring is more expensive and therefore only installed when specified. In contrast to residential wiring, commercial wiring requires the protection of additional layers of insulation and is typically installed within a metal conduit.

Magnetic fields are only generated when current is actively flowing. Although magnetic fields radiate from power lines *at all times* with varying intensity, within the home, electrical wiring, power cords, and appliances only produce magnetic fields *when they are in use*. Magnetic field strength is directly related to current, i.e., amperage. If wiring has been done properly, the current going to an outlet or device will equal the current going back to the power source. When live and neutral wires are in close proximity–and their current is flowing in opposite directions–the magnetic field generated by the current in each wire will largely cancel each other out and result in very small magnetic fields. However, in case of ground faults, where the hot and neutral wires carry different amounts of electricity, a net current will result. This can occur if, for example, a neutral wire from a device is hooked up to a neutral wire from another circuit or the ground wire. A current imbalance will produce a strong magnetic field. Wiring errors such as this are not unusual and can occur in any commercial or residential building.

Magnetic fields are commonly generated by appliances. Devices that draw large currents or contain electric motors, such as toasters or electric fans, produce much higher magnetic fields than others. Hair dryers, blenders, food processors, and even can openers will all transiently expose the operator to magnetic fields while in use. As expected, appliances drawing high current, such as hair dryers, blenders, and drills, generate stronger magnetic fields. Magnetic fields penetrate easily through most materials, and, like electric fields, their strength drops with the square of the distance from the source.

To recap, pure AC electricity generates ELF-EMF. The frequency and wavelength of the EMR emanating from all components of the electrical grid should match the oscillation of current in the live wire and produce a sinusoidal wave with a frequency of 50/60 Hz. The amplitude of the electric and magnetic waves is dependent on the voltage and current, respectively. The radiation intensity drops off dramatically with increasing distance from the wire. That would be the end of the story, except for the presence of voltage contamination on electrical lines, a phenomenon called *dirty electricity*.

Dirty Electricity

The first time I learned about dirty electricity, I was naively unimpressed. I didn't care for the name and the concept seemed too abstract. In my ignorance, I thought the phenomenon was inconsequential. But I was definitely mistaken and uninformed! After the utility company put a smart meter, a notorious generator of dirty electricity, on our home, I became acutely aware of the anxiety and insomnia that this form of electro-pollution can cause.

Very few, if any of us, are exposed at home or at work to pure 50/60-Hz electric and magnetic ELF-EMF emanating from wiring. This is because, in addition to the 60-Hz current, indoor wiring is often contaminated with sporadic, high-frequency voltage transients, and this phenomenon is growing with the increasing use of switching electronic devices, meaning devices that have the ability to open up or close an electrical circuit. The original label for this phenomenon, *electromagnetic interference*, has mostly been replaced by the term *dirty electricity*. This term was coined by Sam Milham, MD, MPH, who authored an excellent book by the same name. Think of dirty electricity as a wastebasket term referring to any stray electrical current that is in an improper location, or has an amplitude or frequency that is anomalous and extraneous to the 50/60 Hz.

Because an electrical line powers all appliances, devices, and bulbs in a dwelling simultaneously, when any one device is turned on or off, it disturbs the line voltage much like water pressure may transiently drop or temperature may flux, for example, in the shower, when a toilet is flushed. In addition, some gadgets and appliances abruptly cut off the incoming electrical current at random segments of the sine wave as they click on and off. By switching currents on and off during segments of the 60-Hz wave, sharp "corners" are introduced to a normally smooth waveform. These corners introduce higher-frequency signals at 120 Hz, 180 Hz, 240 Hz, etc. called *harmonics*, into the electrical grid, which radiate electrical and magnetic fields at those higher frequencies.

Nowadays, this is exceedingly common. Many devices run on voltages other than 120 V, some as low as 3-5 V. Laptop computers, televisions, cell phone chargers, and desktop computers all run on DC power, rather than AC power. Electrical conversion for each device from 120 V AC to lower levels of DC power can be achieved with high efficiency using a switch-mode power supply, typically a charge cord adapter accompanying the product.

In terms of lighting, dimmer switches and some light bulbs, such as CFLs and LEDs, also produce dirty electricity. CFLs contain a ballast that steps up voltage, as these bulbs operate at a very high voltage of 3,000 V. LEDs convert AC to DC with the aid of a switch-mode power supply in their base. In our attempt to improve efficiency and save energy, we have inadvertently exacerbated contamination of our electrical lines and, as we will see, electro-pollution in our indoor environments. Many appliances create dirty electricity.

Common dirty electricity generators

Appliances	Equipment	Electrical systems
Dryers	Desktop computer	Dimmer switches
Washer	Printers	CFLs and LED lights
Microwave ovens	Laptop/cell phone/tablet	Low-voltage lighting
Kitchen appliances	Televisions	Smart utility meters

Personal Energy Generation and Dirty Electricity

In looking to become more energy sovereign, many have opted to produce their own electricity, most commonly with solar cells. Some systems tie into the electric company's power grid, while others are isolated, or "islanded." Islanded solar "microgrids" are usually 48-V DC rather than 240-V, 50- or 60-Hz AC, but these signals still radiate electric and magnetic fields and create dirty electricity, as they still supply power to switch-mode adapters. In addition, DC signals are also regulated by DC-DC converters to levels needed to charge storage batteries on-site. DC-DC converters are switch-mode circuits, and they produce a lot of transients and therefore dirty electricity. Islanded DC power must often be converted to 60-Hz AC to interface with the grid and be compatible with 110/120-V household appliances. To do this, the energy must be converted into AC, using a circuit called an *inverter*. Inverters are also switch-mode devices, and their high current levels make them strong producers of dirty electricity. Finally, generators with switching gear can also produce dirty electricity. All of these switch-mode circuits contribute to dirty electricity and the increasingly distorted electric and magnetic fields that result.

Although wind turbines generate AC electricity, the process is not clean. Because wind speeds fluctuate, blades don't spin at the same speed all the time. The AC power that results must then be converted to DC using a rectifier and then back to 60-Hz AC using an inverter. Both steps involve switch-mode circuits. A wide range of frequencies from below 1 Hz through the kHz range may be observed around wind turbines. Varying currents and voltages are also produced.[65] The prevalence of dirty electricity and proliferation of stray currents from wind turbines can be improved if collection lines are buried.[66] But this will then lead to more ground current.

Dirty Electricity Produces VLF and LF

If damage from dirty electricity were limited to the effect noise has on electrical equipment and computer "crashes," it would only be a dilemma for electrical engineers. But unfortunately, stray currents and voltage transients produce electromagnetic fields. But unlike AC current, these transients are sporadic, vary in frequency and intensity, and are often in the kilohertz frequency range, corresponding to VLF (3-30 kHz) and LF (30-300 kHz). These emissions are not only produced by the appliance itself but also emanate from attached wiring, which can function like an antenna and distribute the anomalous currents throughout the house. The radiation produced by dirty electricity can travel for many feet, depending on the voltage and frequency of each individual transient. I've heard a building biologist describe the detection of VLF from voltage transients that were produced by an appliance on the other side of a house.

Dirty electricity is more than likely being produced in your own home. But in addition to that, it is likely coming into your home from your neighbors' houses and/or local businesses. Because our electric grid connects our houses to one another, in addition to houses and businesses down the road, the contamination of wires in one home or business can travel along wires to other homes. Not only that, in many urban environments, buried copper and iron water pipes have been used as a ground rather than placing unique grounding rods for each house. These same pipes supply water to all the residences and businesses in the neighborhood. If there is faulty wiring anywhere on the block and a neutral wire is connected to ground, which is then connected to a water pipe, stray currents can travel along the water pipes and plumbing systems of other houses on the block. When I first learned about this, I thought it was probably rare. But when a building biologist came to my house and demonstrated that I had stray voltage currents on the plumbing throughout my house, I came to understand it is much more common than I had realized.

In summary, our electrical systems involve much more than the eye can see because wires emit invisible fields of radiation that contain electrical and magnetic components. Exposure is often unpredictable due to anomalous stray currents and voltage transients contaminating our electrical wiring and potentially our plumbing as well. ELF-EMF produced by electricity and VLF and LF from dirty electricity have been shown to have potentially detrimental health effects. In the next chapter, we will explore some of the scientific research behind these claims, and in following chapters look at ways to reduce our exposure and lessen our risk of disease.

CHAPTER 4

HEALTH IMPACTS FROM THE RADIATION PRODUCED BY ELECTRICITY

We live in toxic soup, slowly being irradiated with microwaves and lower-frequency electromagnetic waves. Industrial chemicals and heavy metals contaminate water supplies. Foods are tainted by chemicals used in production, packaging, and preparation for increased shelf life and storage. Air pollution is a source of toxicity in many cities *and* rural areas. Synthetic compounds are damaging our systemic health, including the health of our microbiome. In response to the oxidative stress and other damage many of these toxicants cause, most physicians and wellness advocates promote the benefits of a nutritious diet and exercise to stay healthy. Fewer acknowledge the importance of avoiding environmental toxins. It's not politically advantageous to do so. Emphasizing healthy lifestyle choices is certainly a giant leap forward in the right direction, but the biological effects environmental toxicants have on public health are now and have been profound.

Harmful Health Effects from Electricity?

When most people think about getting hurt from electricity, they think of receiving a big jolt of energy, i.e., electrocution. If we inadvertently complete a circuit between a power source and ground, we effectively become the "load" and electrical current will run through our body to ground, potentially damaging structures along the way. Electrocution can result in unusual paths of injury, depending on the entry and exit sites of the current. Pathways of tissue can die from the burst of excessive energy. The toxic energy load can also be fatal. Regulators are well aware of the inherent risks electrical currents pose. Electrical codes are designed to prevent injury and fire.

Electric shock or electrocution provides an immediate sensation and an "acute traumatic injury." Attributing tissue damage to a lightning strike is straightforward. The cause and effect are easy to see. Chronic injury, occurring from long-term exposure to electrical current, is more nebulous. Although electromagnetic fields produced by alternating current power can produce chronic health effects, proving this relationship is challenging. This difficulty is not unique to EMFs. How long did it take for scientists to prove to the public that cigarette smoking causes cancer? Or even more abstract, how long did it take to prove that secondhand smoke can cause lung problems and heart disease?

Unlike cigarette smoke, electromagnetic energy is invisible and odorless, unable to be sensed. Even when we intellectually understand the science, we may want to believe that our bodies are impervious to this energy. But they are not. Energy waves do not respect the interface between air and skin, or muscle and bone for that matter. Fields emanating from electrical wiring can penetrate through a wall or a human body. The extent to which the radiation reflects off the body's surface, passes entirely through the body, or absorbs into different tissues is one factor that determines its effect. If the radiation does penetrate into the body, the *depth of penetration* will depend on the frequency of the radiation and other factors, including the body's composition and state of hydration.[67] But the scientific term *penetration depth* is a misnomer, as it is not the *limit* of the radiation's penetration, but rather the depth at which the power density of the radiation has dropped to one-half to one-third its original value.

$$\text{Power density at the penetration depth} = \text{power density at the skin surface}/e$$

$$(e = \text{Euler's number, an exponential constant with a value of } {\sim}2.7)$$

A higher power density at the skin surface will result in a greater power density at the penetration depth, as well as in deeper tissues. So, for all intents and purposes, depending on the original intensity of the source, plenty of radiation can be left at the calculated depth of penetration. Furthermore, mechanical deformation of the skin may increase penetration depth. You can imagine that by sitting, for example, with the buttocks flattened and pushed aside, radiation from an electric recliner or massage chair will reach deeper structures than it would if the person received the same exposure to the same area from a standing position.

This concept will benefit from an analogy. Imagine you are driving during a misty night with your headlights on. The headlights illuminate the mist in front of your car, and you can only see for a short distance. Light is absorbed and scattered by the mist. The distance where the brightness of the light reaches about two-thirds the intensity of the original light coming from the headlight is considered the penetration depth. Perhaps it is only thirty feet in a dense fog. If you put on your "brights," the light coming out of the headlamp will be more intense. The technical term for the increased light intensity is increased *power density*. The more intense light will dissipate at the same rate as before, but at the penetration depth, there will be a greater amount of light because the power density of the original light source has increased. The fog will appear much brighter, preventing you from seeing a safe driving distance. It is for this reason we are taught not to use high beams in fog. Understand, though, that even though we can't see for any great distance in this scenario, light projects beyond the penetration depth into the fog. If there was a person in the far distance, you wouldn't be able to see them, but they would still be able to see your headlights because there will be enough light penetrating through the fog to create a biochemical reaction in their eyes.

The Importance of Study Design

Scientists have been studying the health effects of electromagnetic radiation for many decades and publishing thousands of peer-reviewed papers in the scientific literature using various study designs. Two of the more common types of studies used to evaluate the relationship between toxic exposures and people's health are cohort studies and case-control studies.

Cohort studies look at the health outcomes of an exposed population and compare the incidence of diseases to a similar population without this exposure. Given that electricity and electromagnetic radiation are fairly ubiquitous, you can imagine that finding an unexposed control group is very difficult. The Amish in Lancaster County, Pennsylvania, have sometimes been studied as a control population because they do not use electricity.[68] More commonly, however, researchers have focused on assessing the health risks of those with high occupational exposure to electromagnetic fields and compared their health with those in the general population. The thought is that if exposure can cause disease, those with the greatest exposure should have the highest incidence of disease. Electricians, welders, and linemen who repair telephone or power lines are some of the occupations with high routine exposure to ELF-EMF.

Case-control studies differ from cohort studies. In a case-control study, a population with a specific disease is compared to a control group from the general population. Scientists perform this type of study to see whether exposure to a specific condition or toxin is higher for the group with disease. A case-control study might, for example, take history assessments from a group of people with Alzheimer's disease and determine whether they had a higher relative exposure to ELF-EMF historically than a group of individuals of similar age and demographics from the general population that do not have Alzheimer's disease.

As with biochemical toxin experiments, experiments assessing the effects of electromagnetic exposure on animals can point to human health effects. Researchers typically compare the health outcomes of animals placed in different controlled environments, ideally limiting the variables to the one being questioned. For example, a study might divide up one large group of rats, all with the same genetics, into two groups: A and B. Group A would then be exposed to high ELF-EMF whereas group B would serve as controls and receive no exposure. In this way, the researcher can determine whether the exposed group has different health outcomes. Industry often pooh-poohs medical research performed on animals, declaring that effects in animals don't necessarily correlate to human effects. But that's disingenuous, and they know it. Medical research has used animal modeling studies for thousands of years.[69]

Drug companies and personal care product manufacturers consider testing on animal models to be crucial.

Researchers also study the effects of electromagnetic radiation on plants and cell cultures. As in animal studies, one group is exposed while the other is not. The variation in growth and other changes between the two groups can help assess whether the radiation affects the cell's physiology.

ELF-EMF Research Challenges

When evaluating the significance of a study's results, it is important to evaluate details of the protocol and statistical analyses. Reproducibility is key, meaning that in order for research to be considered valid, other labs need to be able to copy the study's design and get the same results. If data are not reproducible, the original researcher's conclusions become suspect. ELF-EMF study protocols have been anything but standardized over the years among labs, and study quality has varied, as is the case for all scientific research. Scientific reviews synthesize the results from a selection of studies over a given time period and report on variability and/or consistency of findings. Meta-analyses are a type of focused, in-depth review that aggregates data published by multiple scientists researching a specific subtopic. The goal of the meta-analysis is to provide an overview of the historical data and determine whether the reported findings of different labs indicate a recognizable trend or association. In this way, the meta-analysis can help establish causality—or not.

In general, reproducibility in this field of scientific research has been lacking. There are a couple of reasons for this. First of all, protocols have varied. Variables such as the duration of exposure, the frequency of the radiation being studied, and the intensity of the exposure can all affect radiation's "bioeffects." A second consideration is that, like ionizing radiation discussed in chapter 2, health effects from ELF-EMF may appear to be seemingly random, "stochastic," and/or predictable, "determin-istic." Deterministic effects, such as heating of tissue, are easier to reproduce. Stochastic effects are hard to reliably demon-strate unless very large populations are studied. Importantly, the stochastic effects can be much more detrimental and lead to chronic disease. A third factor to consider is the underlying health status of study subjects, which can impact whether or not disease occurs. These specifics, and others, account for the variability of findings in the scientific literature.

Intensity and Frequency Windows

Most research on electromagnetic radiation has investigated the health effects of 50/60 Hz, but exposure intensities have varied, as has the duration of exposure. It may seem intuitive that increasing the exposure intensity would cause a more severe effect. Toxicologists, when studying potentially harm-ful chemicals, look for a dose-response relationship between a negative health outcome and the amount of exposure to a biochemical toxin. This relationship is one of the Bradford Hill criteria that helps to prove causation. But in the world of EMFs, a dose-response relationship has not been shown to reliably occur. Instead, *intensity windows* have been demonstrated, in which there are biological reactions that occur within a given range of intensity, while below or above this window there is less, or even no, harmful effect.[70]

A recent study demonstrated an intensity window when studying the effect of electromagnetic radiation on white blood cells in the umbilical cord; when the intensity was between 6 and 13 microtesla (μT), there was an observable decrease in the number of viable cells, indicating that the radiation caused harm in that intensity range. This effect, however, was not seen outside of this intensity window.[71] These response win-dows have complicated research. Changing the duration of exposure can produce opposite responses in cells, such as suppression from one set of parameters and hyperactivation from another. It is important to understand that just because health effects are not consistent across studies with varying protocols, that doesn't mean they aren't real.

Not all electromagnetic field research has been performed at 50/60 Hz. Studies done using other frequencies have shown that *frequency windows* also exist. Similar to the intensity window, these frequency windows define a range of frequencies where physiologic effects occur, and outside of this window, there is no apparent effect. In one example, radiating rats with 50 Hz changed the shape and affected the differentiation of ovarian cells during embryogenesis, which could cause infertility or reduced fertility.[72] But a similar study looking at chronic exposure to 30 Hz did not show a significant difference between the exposed and control groups.[73] The difference in results between these two studies suggests a frequency window that has yet to be fully understood. This makes sense as we experience the various frequency spectra differently with our bodies. For example, most of us can't see the infrared frequency band, but we certainly can feel its presence as heat. We can't see ultraviolet frequencies either, yet we experience their effects as sunburn and an increased production of vitamin D. But the frequency band between infrared and ultraviolet is the visible light spectrum, frequencies we can see but not feel.

Finally, it is important to point out that an author's ties to industry, including funding for a research project, can affect the outcome of the research study. If the funder has a vested interest in the outcome, the results will be almost certainly skewed toward an outcome favorable to that interest.[74] It should be relatively easy to identify biased research by looking at funding sources, which must now be disclosed in every peer-reviewed scientific paper. But such industry-sponsored papers are often included in meta-analyses, where they get lost amidst the independent research data. If the author of the meta-analysis does not account for the likelihood of bias, these studies effectively contaminate and dilute the statistics, leading to results that are inconclusive when they shouldn't be. Study conclusions may therefore be misleading. A researcher may conclude, for example, that there is no definitive association between electromagnetic radiation and a disease or disorder when, in fact, without inclusion of slanted research, the results *would* be conclusive. This uncertainty confuses government leaders and regulating bodies. Deliberately causing similar confusion is how tobacco companies were able to muddy what would otherwise have been clear and compelling evidence of smoking's deadly effects for so long.

The Research

The first peer-reviewed scientific paper associating magnetic fields from power lines with childhood leukemia was published in 1979.[75] A Colorado community living near high-tension power lines experienced a cluster of leukemia cases in children. The prevalence of the disease appeared to be related to high doses of electrical currents. The possible association of this terrible disease with high-tension power lines caused public fear and protests. Three years later, the same authors found an association between adult cancers and high-current electrical lines.[76] But questions remained. How much magnetic field exposure is *too* much, putting people at risk for cancer? And, importantly, if electrical lines cause cancer, what other diseases could they cause?

Although the concern remains, not everyone has a choice of where they can afford to live. While many buyers don't think twice before purchasing a home adjacent to a high-tension power line, others are wary. I was struck by a casual conversation in the spring of 2023, when a colleague told me she knew of a community built near a high-tension power line in southwestern Pennsylvania in which every household had at least one member who had developed cancer. I didn't research the community myself to determine her statement's validity, but I was impressed that the correlation of cancer and living near high-tension power lines continues to circulate, despite the uncertainty as to which power lines are truly dangerous. Many people aware of the potential for health concerns from power lines believe the long-distance transmission lines are the only ones that should be avoided for they emit the strongest magnetic fields. I'm not so sure. In my experience surveying homes, high magnetic fields can also be found within homes near residential lines even though they carry significantly less current.

We know electromagnetic radiation can affect many different organ systems. Neurodegenerative disease, cardiovascular disease, and endocrine disturbances are among a handful of common conditions that historically have become much more prevalent in communities with increasing technology.[77] I remember during medical school, a nephrologist (a kidney doctor) told his class that the kidneys were the most important organ in the body. One thing we all learned early on was that every specialist believes their area of expertise is the most important. Despite the nephrologist's statement, I believe the brain is the most important organ. Unfortunately, most people take their ability to think, remember, and process information for granted. Although many enjoy escaping their reality by getting "high," a properly functioning brain and nervous system is more important than almost everything else.

Neurodegenerative Disease

A large body of research implicates ubiquitous electromagnetic radiation arising from our electrical systems as a toxic form of energy that is damaging brains and contributing to the development of neurodegenerative disease. As previously discussed, EMF exposure can result in the overproduction of reactive oxygen species and lead to oxidative stress in the nervous system, just as it does in other organ systems. Symptoms can include annoying headaches, fatigue, and disturbed sleep, but chronic exposure can also lead to neurodegeneration in some cases.[78] Parkinson's disease, multiple sclerosis, Alzheimer's disease, Lewy body dementia, and amyotrophic lateral sclerosis are some of the serious neurodegenerative diseases that are becoming more common as our environment becomes increasingly saturated with toxins, including ELF-EMF and dirty electricity.

Dementia

Dementia is excessive deterioration of cognitive function over time. Memory loss and impairment in abstract thought are often accompanied by a change in personality. The concept of dementia goes back thousands of years; descriptions of dementia are found in ancient Egyptian and Greco-Roman texts.[79] *King Lear*, the classic written by Shakespeare around 1605, dramatizes the insidious process of dementia as Lear's behavior gets progressively more erratic. Too many of us have watched the decline of loved ones who have suffered or are currently battling dementia. My family lost two beloved members to dementia, one to Alzheimer's disease and the other to Lewy body dementia.

There are many different subtypes of dementia, each classified by their clinical presentation and their patterned impact on the brain. Multi-infarct dementia, a term initially coined in 1974 by Hachinski, is probably one of the oldest known forms of dementia. It is caused when a person suffers multiple small infarcts, or strokes, over time. In this condition, memory deficits slowly accumulate in a stepwise fashion over time.[80] Alcoholics may also suffer memory loss due to progressive damage to the brain caused by excessive alcohol consumption. Prions, tiny infectious proteinaceous agents, can also cause dementia, but are rare. Although these types of dementia have probably existed for thousands of years, they are not the most common forms of dementia in present-day society. Alzheimer's disease and Lewy body dementia are the two most common forms of dementia. And both diseases have unusual microscopic findings with some overlap.

Alzheimer's Disease

Alzheimer's disease, first named in 1910, has been extensively studied. Many theories have been proposed to explain its increasing incidence, yet its cause remains uncertain. In 1985, Daniel Perl observed the presence of aluminum in tangled neurons within the hippocampus of the brain, making a connection between aluminum deposits in the brain and Alzheimer's disease.[81] But how would aluminum get into our bodies, cross the blood-brain barrier, and get deposited into our brains? When it functions normally, this critical barrier, composed of a specialized network of cells, shields the brain's neurons from unwanted toxic chemicals that could otherwise enter from the bloodstream. Aluminum has long been known to cause neuron cell death (apoptosis).[82] Since Perl's discovery, aluminum-free products have surfaced and become preferred for many consumers. Because the brain's neurons transmit electrical current, it seems likely that aluminum in the brain would affect this current and, therefore, the neurotransmission of signals, thereby causing symptoms. You might think that adding radiation exposure to a brain housing aluminum deposits would compound the aluminum's effect, but this has not been demonstrated.[83] Perhaps this is because aluminum is not ferromagnetic (able to be magnetized by a magnetic field) and therefore not affected by magnetic fields.

Many occupational studies document an increased incidence of Alzheimer's in workers routinely exposed to high-strength magnetic fields. A study conducted in 2000 found that seamstresses, dressmakers, and tailors, who routinely use sewing machines that produce strong electromagnetic fields, have four times the risk of developing Alzheimer's than individuals without occupational exposure.[84] Other studies have shown less dramatic results. A study looking at men with long-term occupational exposure found an increased incidence of Alzheimer's disease, but the same result was not found for women.[85] A case-control study evaluating 195 patients with different types of dementia, including multi-infarct dementia, and 229 controls determined there was no conclusive evidence that occupational exposure causes dementia.[86]

I mention this study to illustrate how process and research technique can affect a study's conclusion. This research was less than helpful because the researchers lumped all dementia patients together, without regard to the disease's underlying mechanism. Including patients diagnosed with multi-infarct dementia, where the cause is known, with those who have Alzheimer's in an attempt to find an overarching cause for dementia makes no sense. A potential cause for dementia in Alzheimer's disease could easily be obscured, diluted by those patients with multi-infarct dementia.

Numerous studies have been performed over the years to assess the relationship between electromagnetic radiation and Alzheimer's disease, and a recent review of the literature concluded that *occupational* ELF-EMF exposure may correlate with an increased risk for developing Alzheimer's disease.[87] Yes, the investigators went out on a limb to suggest that occupational exposure could be causing neurodegenerative disease.

Meanwhile, memory centers have popped up all over the country. For those of you who are younger, this may seem like a normal annex to an assisted living community. But is it really? I do not believe the dramatic increase we are seeing in the number of people afflicted with Alzheimer's disease is merely because people are living longer. The incidence of dementia is projected to increase by threefold by 2050,[88] while the elderly population of the United States is projected to be 87 million in 2050, up from 74.6 million in 2020, an increase of less than 17 percent.[89] In other words, the elderly population is not going to triple in twenty-eight years.

Lewy Bodies

Lewy body dementia is associated with abnormal clumps of a protein called $\propto$-synuclein which accumulate in the brain and affect its function. These "Lewy bodies" were initially described in 1912 by Dr. Friedrich Lewy, a German neurologist, who discovered them in patients with Parkinson's disease and suggested a viral cause. In addition to Parkinson's and Lewy body dementia, these protein aggregates are found in a part of the brain called the *amygdala* in a majority of patients with Alzheimer's disease.[90] Multiple system atrophy (MSA), another neuromuscular disorder, is also associated with these abnormal protein aggregates.[91]

The protein, $\propto$-synuclein, is normally found in the pre-synaptic cytoplasm of neurons. But in its normal state it is a monomer, a single-protein molecule. In those with Lewy body dementia, and other Lewy body diseases, this protein undergoes a conformational change. The protein folds upon itself and bonds to other $\propto$-synuclein molecules, forming larger molecules, dimers and oligomers (conglomerations of two or more similar proteins), which are bulky and unmanageable proteins.[92]

Abnormal protein-protein bonds that render proteins inactive are one of the features of denatured proteins. It's likely this protein damage and subsequent bonding to other damaged proteins is the result of oxidative stress. What could be causing this oxidative stress deep in the brain? Normally, the brain is protected from toxic chemicals by the blood-brain barrier, which is very effective. But radiation doesn't respect the blood-brain barrier. Could this protein denaturation be the result of electromagnetic radiation penetrating the skull? It certainly seems plausible.

Some scientists hypothesize that electromagnetic radiation is an environmental cofactor working in conjunction with environmental pollutants like petrochemicals and hormone disruptors to cause a sudden upsurge in neurologic disease.[93] In this scenario, radiation would damage the blood-brain barrier, which would then allow the toxic chemicals to enter the brain and damage its neurons. This seems plausible too. Either way, I suspect ELF-EMF causes the primary damage to the brain.

Neuromuscular Disease

Amyotrophic Lateral Sclerosis (ALS)

ALS is a frightening, progressive neurodegenerative disease in which neurons controlling muscular contraction degenerate and die, leaving the brain unable to control the body's musculature. This disease was first described in 1869 by the French neurologist Jean-Martin Charcot and later became known as Lou Gehrig's disease, in honor of the famous Yankees first baseman who succumbed to the disease in 1941. Over the years, researchers have found aluminum accumulation in the brains of patients with ALS similar to that associated with Alzheimer's. Magnetic fields may induce or exacerbate ALS.[94] A recent review of the scientific literature further supports an association between occupational radiation exposure and an increased risk of developing ALS.[95]

Parkinson's Disease

Parkinson's disease is a chronic, progressive, and debilitating neurodegenerative disease that affects movement. Tremor, dyscoordination, rigidity, and slow movements are common symptoms, which can lead to anxiety, depression, and apathy in sufferers. In Parkinson's, Lewy bodies form in a portion of the midbrain called the *substantia nigra*, the area responsible for producing the neurotransmitter dopamine. Oxidative stress kills neurons in the substantia nigra of Parkinson's patients, affecting the patients' ability to control their movements.[96] Pesticides, prior head trauma, and exposure to the dry-cleaning solvent trichloroethylene[97] are some of the predisposing factors that have been associated with developing Parkinson's disease.

Although ELF-EMF seems a likely causal agent, a recent meta-analysis evaluating eleven prior scientific studies failed to find a link between occupational exposure and the development of Parkinson's disease. The authors concluded that electrical workers and others exposed to high levels of radiation are not at increased risk of developing Parkinson's.[98] But, as with all research, the devil is in the details. Most of the studies in this meta-analysis reviewed death records mentioning Parkinson's. Unfortunately, this condition is not likely to be listed as a cause of death, on death certificates of most people who suffered with the disease, which would mean that most cases were missed. This introduces a bias known as *under-ascertainment*. Given the increasing incidence of this neuro-degenerative disease, I believe continued research is needed to determine whether there is a relationship between electro-magnetic radiation exposure and Parkinson's disease.

Multiple Sclerosis

Multiple sclerosis is a neurologic disease with a variety of clinical presentations and a variable course. It is a chronic disease but not necessarily progressive in all patients. The relationship between nonionizing radiation and multiple sclerosis is ambiguous. On one hand, dirty electricity exposure produced by electronic equipment and sleeping with an active cell phone or laptop under one's pillow, have been implicated in increasing risk of developing multiple sclerosis.[99] But studies also indicate that patients with multiple sclerosis may see an improvement in their symptom severity with EMF exposure. The difference in effect is likely due to the characteristics or quality of the radiation. A therapeutic application is achieved by exposing the brain to repetitive unmodulated pulses of frequency, called continuous waves, whereas dirty electricity and the radiation coming from a wireless communication device is erratic and not continuous, like a flickering fluorescent light bulb. This will be discussed in greater detail in chapter 13.

ELF-EMF Effects on the Blood and the Immune System

Unlike cells in a stationary organ, blood and immune cells circulate through the body, flowing through tiny capillaries under the skin as well as suffusing the deep organ systems. *Any* penetration of nonionizing radiation can therefore potentially affect blood and immune cells.

After Wertheimer's initial bombshell research in 1979 associating high-tension power lines with childhood leukemia, scientists performed further research in the late 1990s and 2000s exploring this claim. Although some research papers failed to confirm an association between radiation and leukemia, pooled analyses of studies concluded that children who lived in homes with ambient magnetic fields stronger than 400 nanotesla (nT; 0.004 gauss) were twice as likely (relative risk of 2.0) to develop leukemia than children whose bedrooms had magnetic field levels below 100 nT.[100] Most homes have magnetic fields somewhere between 10 and 100 nT. Another meta-analysis performed by Greenland et al. showed a 70 percent increased risk (relative risk of 1.7) in children with household exposures of 300 nT or higher.[101]

A literature review evaluating studies from 2000 to 2019 concluded that magnetic fields associated with ELF-EMF not only put children at risk for leukemia, but also put adults at risk for leukemia, breast cancer, and brain cancer.[102] Although we will discuss detection and remediation in a future chapter, consider 300 nT, or 0.003 gauss, the lower limit of the danger zone for your home and office space. Living spaces should ideally have a significantly lower magnetic field strength.

Aside from the risk of developing cancer, research has shown that EMFs can have a variable effect on the immune system. Electromagnetic radiation can either excite or suppress the immune system, depending on the duration of exposure. The immune system has many different components and cell types that allow it to function. *Macrophages*, an important type of white blood cell, behave abnormally when exposed to radiation. When macrophages are exposed to 27 Hz for thirty minutes, they become less aggressive than normal according to one study. But after sixty minutes of exposure, the macrophages become *overly* aggressive.[103] Overactivity or underactivity of macrophages can significantly affect a person's overall health status. A review article studying the effects of electromagnetic fields on immune cells–in particular monocytes, T lymphocytes, B lymphocytes, natural killer (NK) cells, and macrophages–indicated that radiation can affect the number and function of immune cell lines, including cell proportion, cell cycle, apoptosis, killing activity, cytokine contents, and others.[104] One mechanism proposed for these effects is that an influx of intracellular calcium, instigated by radiation, could result in decreased calcineurin levels. Calcineurin suppression leads to immunosuppression.[105] However, a review focusing on the immunomodulatory effect of EMFs of varying frequencies suggested that explaining the various effects reported by different researchers would require multiple mechanisms which have yet to be discovered.[106]

EMF and the Heart

The heart is a big muscle that repetitively contracts, pumping blood throughout the body. The heart is composed of four chambers: two atria and two ventricles. Synchronized contractions of first the atria and then the ventricles drive the blood forward. If the synchronization is lost, the heart will not effectively deliver blood to the body. This can lead to congestive heart failure, fainting, shortness of breath, fatigue, and even organ failure. If blood stagnates in the heart, it can clot. Blood clots in the heart can then be pushed out into the arteries and damage organs, including the brain, causing a stroke.

Coordinated contraction of the heart muscle is achieved via an electrical signal that travels along a pathway that starts in the SA node of the right atrium. This impulse is then transmitted to another node, the AV node, and then to the rest of the heart. If anything disrupts this signal conduction pathway, an abnormal heart rhythm (*arrhythmia*) can develop.

An interesting study performed on dogs showed that exposure to magnetic fields from extremely low-frequency radiation (2.87 µG at 0.043 Hz) can affect electrical conduction in the heart, heart rate, and rhythm, potentially causing atrial fibrillation.[107] Although the frequency studied is very low, nearly direct current, there may be higher-frequency windows that have the same or similar effects. As with most of this material, more research is needed. Although research has not yet proven an effect on heart conduction or the development of cardiac arrhythmia from ELF-EMF, I am suspicious. I have observed an increase in the prevalence of atrial fibrillation in my own practice as I witness an increasing number of young people below 40 years of age with pacemakers including wireless pacemakers on their chest X-rays. Indeed, the increasing number of people with atrial fibrillation is being referred to as an epidemic.[108] I suspect the heart's conduction system is being influenced by electromagnetic radiation of one frequency band or another.

In medicine, when a patient comes in with symptoms, the goal of the clinician is to find a disease or condition that explains the symptoms. Unless an astute clinician is aware of the myriad of health effects that can occur from environmental toxins, they may compartmentalize each symptom as a separate diagnosis. This is certainly the case for EMF toxicity. Seemingly unrelated symptoms can appear in different organ systems due to electro-pollution. I have personal experience in this as I believe it is what led to my father's demise.

Sickness in Paradise

It had always been my father's dream to retire to the affluent community of Incline Village along the north shore of Lake Tahoe. After achieving his dream and living in his custom-built house for several years, he started calling me fairly frequently to discuss his wife's deteriorating health. She was experiencing a progressing set of strange and nonspecific symptoms, such as ringing in the ears, dizziness, vertigo, and fatigue, which eventually prompted me to visit their home with my EMF detectors. While surveying their home, I found high levels of RF radiation. In addition, my father's office, the room he spent most of his time in, was very close to a high-voltage power line and a large transformer on a utility pole. He bathed in elevated magnetic fields when seated at his desk.

I expressed my concern about the elevated levels of radiation in their home and suggested they start taking steps to try and reduce their exposure. Being an engineer who had worked on radar applications at Raytheon, my father did his own survey and decided that the EMF levels in their home could not possibly be harmful. He was a healthy seventy-something-year-old who taught ski school, walked every day, and had no habits one could consider unhealthy. Within the following four years, he was diagnosed with an arrhythmia and underwent a cardiac ablation. Soon after, he developed dementia and subsequently died. His younger brother, who is now eighty-nine, is still alive, has no signs of arrhythmia, and is mentally "as sharp as a tack." In fact, no one in my father's extended family had ever been diagnosed with an arrhythmia or dementia before. This is an anecdote, of course, but one that raised my level of suspicion.

EMF and Amyloidosis

Amyloidosis is a debilitating disease that can affect many organ systems, including the heart, brain, kidneys, joints, and others. A radiation-induced intracellular calcium imbalance may be a trigger for the production of amyloid-ß.[109] We know EMFs increase the tendency for amyloid peptides to aggregate and form fibrils. The accumulated aggregation of amyloid peptides has been shown to be dependent on the intensity and frequency of the radiation.[110] Perhaps amyloidosis is also a disease resulting from protein denaturation secondary to oxidative stress? Unless you know a person, who has been diagnosed with amyloidosis, you have probably not noticed that amyloidosis centers are opening all over the country. Amyloidosis was considered very rare thirty years ago. But now, there are four different known subtypes and enough patients to sustain treatment centers nationwide!

Health Effects of Dirty Electricity

Studies on the health effects of dirty electricity and VLF are less prevalent than the body of research performed on ELF-EMF. In *Dirty Electricity*, Sam Milham, MD, PhD, describes his research study of La Quinta Middle School 2005-06 investigating a cancer cluster that was initially discounted by a "specialist." In this school, sixteen teachers developed eighteen cancers during the research period, which was 3.2 times the expected rate found in the community. In other words, only 4.7 cancers should have developed in this group of teachers (which is still high!).

Interestingly, four of these teachers developed malignant melanoma, eleven times the expected rate. Rates of thyroid cancer, uterine cancer, and polycythemia vera, a blood disorder, were also excessive.[111] In the unlikely event that the school's teachers happened to have a randomly high rate of cancer, the distribution pattern would almost certainly have been more in line with background rates. Milham detected "off-the-chart" levels of dirty electricity during his survey of the school.

Interestingly, in the La Quinta Middle School, one of the teachers complained that her students were unteachable. The outlets in her room all demonstrated excessive dirty electricity. After remediating the situation by plugging dirty electricity filters into the classroom outlets, the teacher reported that her students became much more attentive.[112] Additional studies have also reinforced the connection between reducing dirty electricity levels in schools and improvements in teacher health and student behavior in middle and elementary schools. The same effect was not observed, however, in high school students.[113] The age discrepancy could be related to radiation absorption differences related to physical size. Alternatively, the difference could be explained by high school students moving from class to class, where they are relieved of intense exposure during brief periods throughout the day, versus younger children remaining in one classroom where they receive constant exposure to dirty electricity and now radiofrequency radiation throughout the day.

Dirty electricity can also elevate blood sugar levels in people with diabetes and prediabetes.[114] Given the increasing prevalence of dirty electricity in our homes and office spaces, it certainly seems plausible that this form of indoor electro-pollution may be contributing to the diabetes and obesity epidemics.[115]

Electromagnetic radiation has likely affected human health for over a century. But the recent development of more complex electrical devices creating dirty electricity has likely exacerbated the incidence of many chronic diseases. My suspicion is that the ever-increasing currents of dirty electricity and their associated transient and erratic VLF and LF emissions may be one of the reasons conditions such as neurodegenerative diseases and cardiac arrhythmias are increasing to epidemic proportions. We have ascribed these diseases to lifestyle choices, such as eating processed foods and lack of exercise. And, while these choices certainly play a role in health, we should not ignore the effect our electrified homes and businesses have on our health.

I strongly suggest we all take steps to lessen our daily exposure to electromagnetic radiation and dirty electricity. This will be a challenging task given the societal push for more and more electrically powered gadgets. In some airports now, rows of seats on concourses even house outlets at head level (Figure 4.1). The electric field levels generated are certainly strong enough and close enough to penetrate the brain. My hope is that as awareness of the detrimental health effects of radiating the brain increases, future building codes will rule out designs such as these plugged-in seats.

Figure 4.2. Electrified seats in an airport

CHAPTER 5

DETECTING AND REDUCING RADIATION EXPOSURE FROM ELECTRICITY

The idea of reducing the amount of something you can't see, hear, or feel may seem silly or even ludicrous. Because our senses can't detect electromagnetic radiation of lower frequencies than visible light, it is easy to dismiss electro-pollution as unimportant. Skepticism is further supported by the disinterest in the topic exhibited by electricians and engineers, most of whom have not learned about the health effects of ELF-EMF. As previously mentioned, electrical codes are designed to reduce the risk of electrical shocks and fires. Therefore, at least for the time being, the onus of creating an electrically safe home is on each of us. In this chapter, I cover ways to measure and mitigate ELF-EMF exposure.

The first move toward assessing your daily exposure to ELF-EMF is to buy a suitable detector. Many styles are available for sale. In general, higher-priced detectors offer a superior quality product, each with different features and sensitivity. More expensive detectors will survey a larger frequency range and may offer a three-dimensional survey of magnetic fields rather than a single-axis measurement. For most people, a less expensive detector is adequate for home and office surveys.

Because the magnetic and electric field strengths are independent from one another at ELF frequencies, it is necessary to measure each field separately. Detectors will typically offer a toggle switch to alternate between electric field and magnetic field measurements. Electric fields are typically measured in volts per meter (V/m). Some vendors produce detectors that report magnetic field strength in milligauss (mG), whereas others use the unit tesla (T). Higher-end units with higher sensitivity provide values in the nanotesla (nT) range (billionth of a tesla), while more crude monitors provide data in microteslas (μT; millionth of a tesla). The conversion from nanotesla to microtesla and milligauss is:

$$1 \text{ microtesla } (\mu T) = 1{,}000 \text{ nanotesla } (nT) = 10 \text{ milligauss } (mG)$$

Tools for Surveying EMFs

Many vendors sell EMF detectors. I have no ties to any particular brand. In discussing specific vendors and models, it is not my intention to direct you to a particular product, but rather to draw your attention to the types of features and specifications to consider when comparing machines.

Cornet, Trifield, and Gigahertz Solutions produce popular EMF meters. Each has several different levels of sophistication. For example, the Trifield model TF2 will survey a relatively narrow bandwidth of ELF-EMF from 40 Hz to 100 kHz with an intensity range of 1-1,000 V/m for electric fields and

0.1-100.0 mG for magnetic fields. This meter will be able to detect 50/60 Hz frequency and a portion of the dirty electricity range. The Trifield meter also has the ability to measure RF radiation, which will be discussed later on.

In comparison, the Gigahertz Solutions ME3830B meter detects frequencies between 16 Hz and 100 kHz. This meter, therefore, will detect lower frequencies of ELF than the TF2. This is the detector that I use. A higher-end unit, the ME8040B, also offered by Gigahertz solutions, will detect even lower frequencies, from 5 Hz to 100 kHz. Is this necessary for the average home surveyor? Probably not. In comparison, a professional meter sold by Gigahertz Solutions, model NFA1000, will detect frequencies within a range from 5 Hz to 1 MHz and will provide three-dimensional surveying of magnetic fields. This is an excellent detector, but probably more than most need. In general, I would recommend purchasing the best detector you can considering your budget. With technology, you typically get what you pay for. Plan to spend somewhere between $160 and $300.

Detectors within this price range from a reputable company will be useful and accurate enough, but these lower-priced meters will not provide a detailed analysis of the specific intensity for each individual frequency within the sensitivity range. This type of report is called a *spectral analysis.* Rather, the meters I have mentioned, and others of similar design, provide a composite reading for all of the intensities within the detector's range of detection. So, if your meter is picking up an electric field measurement of 10 V/m, you won't know for certain whether the emission is coming from 50/60-Hz, dirty electricity, a combination of both, or another source. Common sense will help guide you. Detectors that provide a spectral analysis are much more costly and not worth the expense for the average citizen scientist. If in your surveying you encounter unusual exposures not easily understood, it may be worth it to hire a building biologist to help sort things out. These professionals typically have much higher-grade equipment.

Measuring Electric Fields

After you purchase an ELF-EMF detector, play around with it for a few days. Walk through your home and office space with the detector, taking note of the readings you encounter. Then, make a diagram of your home and select specific locations to survey. Choose areas in which you spend longer periods of time. On the diagram, specify your distance from various outlets, appliances, and electrical cords. Appliances only need to be plugged in and need not be turned on to measure the electrical field they generate. Because the intensity of ELF decreases exponentially with distance, it is important to document where you are when you take measurements. If you measure the ELF coming from an electrical outlet from a two-foot distance on one day, but then measure it from a three-foot distance the following day, you will get a significantly lower value on the second reading.

Record three survey readings from each area of your home in which you spend most of your time. One measurement should correspond to the location of your head, at eye level. The second measurement should be in the lower chest area, where your heart lies, and the third level should be in the area of your lap or waist. If you sit at your desk chair for prolonged periods of time during the day, take these three measurements as if you were sitting in the seat in a working position. During your home survey, be sure to measure the electric field intensity over your bed, particularly where your head usually lies, where your heart would be, and where your pelvis would be (see appendix B for a sample chart).

If you spend time during the day sitting or lying on furniture that "plugs in," be sure to take measurements of those too. In addition to waterbeds and self-elevating beds, couches and chairs are now commonly fitted with electrical controls and sometimes an intrinsic electrical outlet. Record all measurements on your chart. If one or more measurements are high, try to figure out its source. Is it from wiring? Is it from an appliance? Does the reading drop off if you unplug the device, extension cord, chair, or bed?

After you write down your survey data, study them. Although more intense exposures may be transiently encountered as you move through other areas of your home, focus on the areas in which you spend the majority of your time. If you can clean these up, the rest will be much less impactful on your health. This report provides a guide to a component of your daily exposure. I specify "a component" because, as we will learn in the upcoming chapters, the modulation of microwave radiation–the kind that makes our cell phones work–is another major contributor to electropollution.

Remediation of Electric Fields

Depending on which source you read, recommendations for maximum electric field exposure vary. Some recommend less than 6 V/m for maximum electric field exposure for any significant period of time. But according to researchers at the Building Biology Institute, ambient electric fields within the home should ideally be below 0.3 V/m, and below 10 mV/m (.01 V/m) in sleeping areas. Depending on your home, this may be difficult to achieve. Electrical field strengths below 1.5 V/m and 100 mV for a sleeping area is acceptable. If you can keep your electric field exposure below those limits, you are in terrific shape. If not, try to lower them as much as possible. My recommendation is to first concentrate on limiting the electric field strength you are exposed to while you sleep. If the chair you sit in for hours at night or the head of your bed has levels over 5 V/m, remediation should be a top priority.

There are several ways you can reduce your electrical field exposure. The easiest remedy is to unplug extension cords, power cords, and appliances when not in use. If elevated electric fields are emanating from an appliance and/or its cord, unplugging the unit when it is not being used will instantly solve the problem. If you want to keep the appliance plugged in, consider moving the appliance, or the chair, couch, or bed, so you are no longer near the electric fields when you are trying to relax.

In order to reduce electric field exposure while in bed, check to see if there is an electrical outlet behind the head of the bed. If your bed has a deep headboard or lots of pillows, you may be positioned far enough away from the wall that you are not being exposed to elevated electric fields when lying down. If not, you may need to move the head of your bed away from a wall. If there is an extension cord running behind the bed, consider plugging lamps and/or your clock into an outlet further away from the bed. Ideally, try to create an electrically clean environment for your sleep sanctuary.

Some people are so sensitive to EMFs that the electrical fields coming through wiring in the wall can bother them, particularly if dirty electricity is prevalent. There are a few potential remedies to address this problem. The first is to apply an EMF-shielding paint along the lower third of the wall. A product called Y-Shield HSF54 paint is effective at blocking electric fields and works well. One coat of paint will reduce the radiation intensity by 44 dB, or by 99.996 percent! This paint is black and conductive, and therefore needs to be grounded to ensure that no one is accidentally electrocuted or shocked should a live wire come into contact with the painted surface. After the wall is covered with shielding paint, it can be covered over with the normal wall color to conceal it. Another option is to install a "kill switch" to turn off the power to your bedroom before you go to sleep at night. This will effectively remove all electricity from the bedroom wiring and therefore no ELF-EMF will irradiate from wiring into the room. If you have a room in your home that shares a common wall with the bedroom and that other room is on a different circuit, you may need to add a kill switch to that room too, particularly if your bed has been placed against the common wall. A less expensive option could be to move the bed to a different wall.

Kill switches are becoming much more popular, in part because of the increasing prevalence of electrohypersensitivity (EHS), a condition discussed in depth in chapter 12. Although I do not personally have experience with a kill switch, I have been told they are effective. Installing one would obviously mean you would no longer have access to any electronic devices, such as lamps, plug-in clocks, or a TV, in your bedroom after you flip the switch. A kill switch resembles a simple light switch that may be placed in the wall when you first enter a room. Although you might think flipping a circuit breaker would achieve the same result, it does not. Circuit breakers disconnect the hot wire, but leave the neutral intact. If there is dirty electricity on the neutral wiring in the home, it will travel along the neutral wire into the room and plugged-in lamps, etc., and emit electromagnetic frequencies *even if the circuit breaker is turned off.* For this reason, it is best to install a kill switch if you are electrically sensitive. Battery-operated clocks and battery-operated motion-activated lights can be placed in the room so you have access to the time and light, if needed, after the kill switch has been implemented.

If you are building a new home or redoing the wiring in an old home, that is an excellent opportunity to take steps to minimize ambient ELF-EMF within the living spaces, particularly if you are electrically sensitive. Changing the wiring will be a more expensive proposition, depending on the methods you choose, but your relaxation and comfort level in the home will be worth the extra expense. In order to reduce electric fields, you can cover walls with aluminum foil prior to applying wallboard. The foil will shield electric fields emanating from wiring behind its surface. Another option is to ensure wires are installed within steel conduit (steel tubing used to protect electrical wiring). This is an expensive alternative to the more common Romex cabling, but the conduit will effectively shield both electric and magnetic fields.

Surveying Magnetic Fields

Surveying magnetic fields can be a little trickier than electric fields. Magnetic fields are three-dimensional and directional; therefore, the field strength is related to the orientation of the detector as you survey. The meter you purchase to survey electric fields will more than likely also be able to measure magnetic fields. The detectable bandwidth will be the same for both fields, meaning that if your meter can detect electric fields between 5 Hz and 100 kHz, it will also be able to survey magnetic fields between 5 Hz and 100 kHz. But less expensive detectors contain what's called a *single-axis directional antenna*. This means the antenna will only measure magnetic fields in one direction at a time. Surveying three dimensions will require changing the orientation of the meter in multiple, taking time to pause with each orientation so the display will accurately reflect the field strength in a given location. Magnetic fields are oriented at 90 degrees off axis to the direction of current and to the electric field (Figure 5.1). This directionality is not a big deal. It just means surveying will take a little extra time to incorporate changing the detector orientation when using a single-axis directional antenna. More expensive meters are able to measure magnetic fields in three dimensions at once, but even with these more sophisticated machines, some maneuvering is still required.

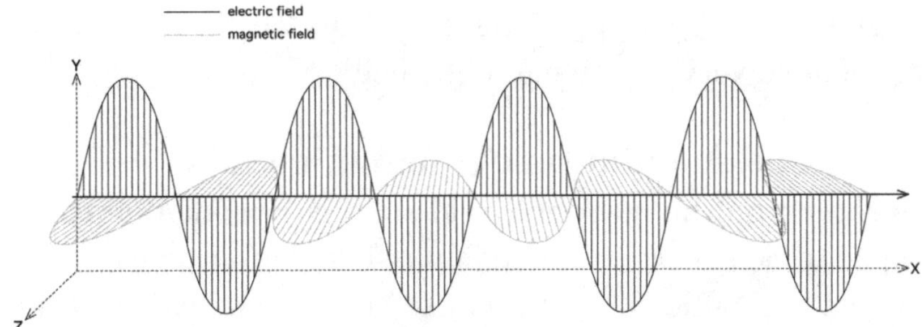

Figure 5.1. Electromagnetic fields

Magnetic field detectors will usually display a number in units of μT, nT, or mG. Revisit the chart you created for electric fields and add a column for magnetic field survey data. Unlike electric fields, magnetic fields are only produced when current is flowing. If, for example, you are lying in bed and there is a lamp with a cord running behind the bed, no magnetic field will be present when the lamp is turned off, but if wiring isn't balanced, a magnetic field will be produced when that lamp is turned on. Perform the survey with each appliance turned both on and then off to determine what, if any, magnetic field is being produced from each. When surveying for magnetic fields, if you encounter significant fluctuation, the highest reading is the one you should enter into your chart in each location (see appendix C for a sample chart).

During your survey, you may encounter an unsuspected source of magnetic fields. When I first surveyed my previous home, I had no idea that I had been sleeping in a very strong magnetic field. The source was an inverter that had been attached to the basement wall in a room beneath the bedroom. The inverter had been installed as part of a solar power system and allowed me to live off the grid by converting the DC power generated by the solar panels and battery backup storage into AC power. At the time, I didn't realize that not only was the inverter creating a large amount of dirty electricity, it was also producing a very strong magnetic field.

Remediation of Magnetic Fields

Depending on the magnetic field source, remediation can be simple, tricky, or pretty much impossible. Most homes have an ambient magnetic field strength somewhere between 0.2 and 1.0 mG (20-100 nT). But it is not uncommon to find a home that has ambient magnetic field levels over 300 nT. According to the Building Biology Institute, we should strive for a magnetic field intensity of less than 100 nT, or less than 1 mG in sleeping areas.[116] The BioInitiative working group set a target level a little higher, between 1 mG and 2 mG (100-200 nT).[117] In comparison, the International Commission on Non-Ionizing Radiation Protection (ICNIRP) guidelines recommend limiting magnetic field exposure for the general public to 400 mT, or 400,000 nT![118]

How could this commission be so out of sync with the Building Biology Institute and the BioIniative Working Group? In their decision-making process, ICNIRP acknowledged that epidemiological studies have *suggested* that long-term exposure to 50- to 60-Hz magnetic fields *might be* associated with an increased risk of childhood leukemia. But because of selection bias and because no biophysical mechanism had been identified to explain the association, this group did not conclude that prolonged exposure to ELF-EMF could cause childhood leukemia or any other cancer. If I told you that some members of this group had financial ties to the industry, I'm sure you wouldn't be surprised. Unfortunately, the World Health Organization (WHO) used ICNIRP's exposure limit guidelines to set safety standards worldwide. Regardless of their rationale, my suggestion would be to stick with the recommendations of nonpolitical groups such as the Building Biology Institute and the Bioinitiative Working Group.

Remediation of magnetic fields may be as simple as turning off an appliance or moving it farther away from a chair or couch. Clock radios typically produce very strong magnetic fields, and can be effortlessly moved farther away from the bed to reduce your magnetic field exposure while sleeping. Magnetic fields dissipate with increasing distance, but they readily pass through building materials. If the source of a strong magnetic field in your home is from outside the house, such as a nearby power line, remediation is not going to be practical.

Thankfully, because of research on the detrimental health effects of strong EMFs associated with high-voltage transmission lines, these dangerous wires are placed high above ground within rights-of-way typically between 75 and 200 feet in width, which is wide enough to protect the public from any harm. Ha ha . . . Just kidding. Although it's true that high-voltage transmission lines are placed in rights-of-way between 75 and 200 feet wide, this width is not based on protecting humans from magnetic field exposure. It is instead to ensure that rights-of-way are wide enough so that uninsulated transmission lines do not come close enough to trees and other foliage to cause an electric arc and start a fire. If you live near a high-tension line, survey to see what you are being exposed to in the home.

If you are considering moving into a house or apartment near a high-tension line, consider assessing magnetic fields to be part of your due diligence before signing paperwork. Survey for magnetic fields even if the utility lines to your home are buried because magnetic fields will travel underground and may expose parts of your home to higher levels than if the wiring were on an overhead line thirty feet above street level! The only way to know is to survey.

Keep in mind that your measurement of the magnetic field strength coming from a power line represents a blip in time. Take surveys at different times of day and perhaps over a year. As demand for electricity waxes and wanes with the seasons and the time of day, the current being delivered to the community will change. So, if you measure high levels in the middle of winter, during a time when everyone in your community has their lights on and their heating systems on, the magnetic field you survey will most likely be much higher than during a spring evening, when the sun is in the sky till later at night and there is no need for air-conditioning or heat.

Be sure to survey at night in the area where you sleep. The magnetic field strength at night, when the neighbors' lights and television sets are turned off, will be less than if you survey during the early evening. If the levels remain at an unhealthy level–over 3.0 mG (300 nT)–and there are no other options for you to lower your exposure, particularly while sleeping, you may want to consider moving.

If you are constructing a new home or thinking of rewiring your old home, have the electrician twist the wire cabling to be installed. Twisting the hot and neutral wires around each other will reduce the magnetic field strength. The more twists, the greater the cancellation effect.[119] Alternatively, install pre-twisted electrical line. As mentioned earlier, installing wires in steel conduit will also shield both electric and magnetic fields.

Electric Seat Warmers and Heating Pads

Electric blankets, heating pads, and waterbeds are just a few of the many electric products designed to produce heat. Battery-powered devices running on DC will produce a static magnetic field. AC-powered devices produce EMFs. The health risk from occasional use of these products is probably minimal. The only study I could find assessing the safety of these products was performed in 1999 and concluded that neither electric blankets nor electric waterbeds were associated with an increased prostate cancer risk.[120] Regardless, I do not recommend applying *any* electrically powered heating unit directly to the skin for prolonged periods of time. If you need the comfort of a heating pad, the old-fashioned rubber variety is a better "go to." Although hot water bladders may be more cumbersome than electric heating pads, I find the heat produced from them much more relaxing and I suspect they are a healthier alternative. Electric heating pads generate very strong cyclical magnetic fields. Why subject the pelvic organs–including the ovaries, cervix, and uterus in women and the prostate in men–to strong magnetic fields if you don't have to?

Electric seat warmers in automobiles run off the car battery. If your vehicle doesn't have this feature, you can purchase an electric seat cover that will plug into the car's power outlet. Because these amenities run on DC power, they are probably relatively safe, but I personally do not use them. Prostate, anal, rectal, and gynecological cancers are much too common, and although their causes are likely multifactorial, it's possible that repeated long-term exposure to a strong static magnetic field while sitting on a contoured seat could be dangerous. I am not aware of any research done to date to confirm or disprove this possibility, but just knowing that DC electricity generates a static magnetic field is enough to keep me from using an electric seat warmer. If you live in a cold climate and don't want your tush to get cold, cover the seat in the wintertime with heavy fabric or another alternative.

Surveying Dirty Electricity

Surveying for dirty electricity is less straightforward and more ambiguous than for ELF-EMF. These stray currents can easily be seen with an oscilloscope agitating the normally smooth contour of a 50/60-Hz sinusoidal wave form of an electrical power line (Figure 5.2).

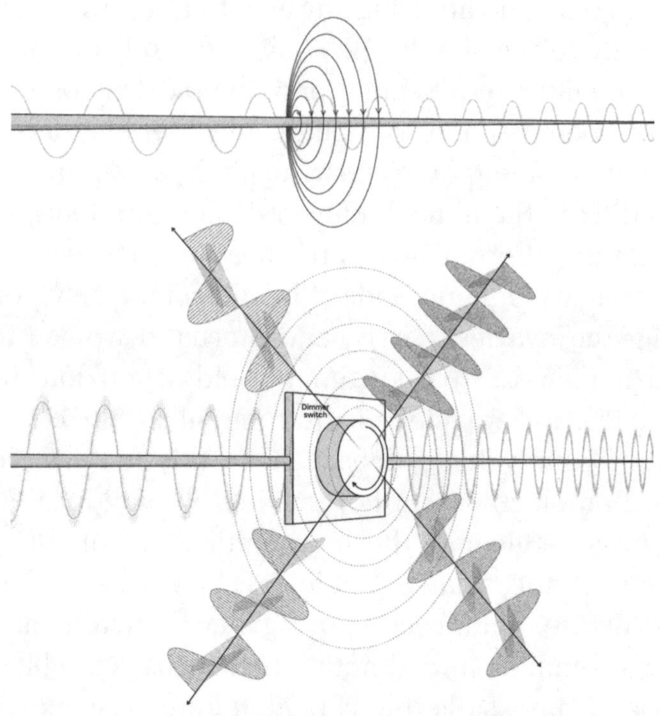

Figure 5.2. Dirty electricity

The upper image displays a normal smooth sine wave associated with clean electrical current and the production of a magnetic field. The bottom display shows the impact of a dimmer switch creating line distortion and production of electromagnetic fields.

Given the prevalence of dimmer switches, CFLs, LEDs, appliances that manipulate voltage, and smart meters, high-voltage transients commonly contaminate home wiring. Because there are typically multiple sources of dirty electricity, the frequencies are variable and may be within the ELF-EMF range or the VLF/LF range, from 24 to 100 kHz, as defined by the WHO. Before surveying, there are several steps you can take to reduce dirty electricity in the home. Remove dimmer switches and install regular switches. Replace LED lights and CFLs with incandescent bulbs. Unplug computer, laptop, and cell phone charging cords when not in use.

Unlike ELF-EMF, surveying for very low frequencies (VLF/LF) is usually done indirectly by evaluating the electrical lines for dirty electricity. Two well-known companies produce plug-in meters that will measure the level of voltage transients on the home wiring, Greenwave and Graham-Stetzer. The Greenwave meter can detect electrical frequencies from 3 kHz to 10 MHz. The Graham-Stetzer meter can, according to their website, detect line harmonics and low-level high-frequency transients from 180 Hz into the kHz frequency range. Both devices plug into ordinary electrical outlets and provide vendor-specific numerical values that represent the intensity of dirty electricity in that outlet. These meters do not provide information about the source of dirty electricity or the frequencies being picked up. They merely provide a ballpark of how bad the contamination is on the electrical line.

In order to detect individual sources of VLF/LF, your ELF-EMF meter, depending on which you choose, may be able to pick up on some of the frequencies produced from dirty electricity. Another option is to use a handheld battery-powered AM radio, which can be used as an inexpensive method to detect sources of VLF/LF in the home. AM radios will pick up electromagnetic interference in the dirty electricity range. Turn on the radio and tune in to any station between 800 and 880 kHz in an area of your home where you can pick up the station without static. Then, walk around the house to various appliances, switches, etc. You will hear a loud buzzing sound when you approach a gadget producing dirty electricity and VLF/LF. If you unplug the device and the buzzing goes away, that's immediate confirmation that a source of dirty electricity was discovered and successfully remediated.

Surveying for dirty electricity or for VLF/LF will help somewhat, but truthfully, if you live in a community, your electrical lines are likely contaminated by a combination of voltage transients originating from appliances in your own home as well as those in the neighbors' homes and businesses down the street. In addition, it's more than likely that smart meters have been installed on the exterior of your home by utilities, including electric, gas, and water companies. These meters produce extensive dirty electricity in addition to MW radiation. Dirty electricity produced by a smart meter will travel throughout your home wiring. It is getting more and more difficult to request (or demand) that utility companies remove their smart meters from the facade of a home. I know some people who have been successful and have had utility companies place the meter on a post away from the house and closer to the street, but most often I hear that there is nothing the utility will do. And there is nothing that you, the homeowner, can do to force them to remove the meter. Gratefully, there are alternatives to "clean" dirty electricity from your home's electrical lines.

Remediation of Dirty Electricity

Cleaning dirty electricity from the lines in a house can be achieved in a few different ways. First off, both the Greenwave and Graham-Stetzer companies sell filters that "clean up" the stray voltages, reducing the dirty electricity coming from the treated outlet. Other outlets, however, will still emit dirty electricity and may similarly need to be treated, depending on where you spend most of your time while in the home. If you plug in dirty electricity filters, and resurvey with a Graham-Stetzer meter, dirty electricity levels should be brought down to 50 Graham-Stetzer (GS) units or less. If surveying with a Greenwave meter, levels should be below 25 mV.

Rather than installing individual filters on the outlets in each room, several companies have designed whole-house filters. An electrician needs to install a whole-house filter at the circuit breaker panel. Discuss your home electrical system with the company you choose to make sure you get the proper product for your load. I purchased a device called the SineTamer, sold by Power EMT, which I have been very happy with. An important advantage to this device is that it removes stray voltage from the live wires, in addition to the neutral and ground wires. The vendor claims this filter will cause a drop in the utility bill. But as I see it, this is secondary in importance to the health benefit from eliminating VLF/LF from the home. As soon as our SineTamer was installed, my children and I felt a subtle sense of relief. It was a funny moment. The three of us were sitting on the living room couch and, at the same time, all looked at each other with a curious expression on our faces. We had simultaneously relaxed. Then, the electrician came upstairs from the basement and announced that the unit had been installed.

There are other companies that produce and sell whole-house dirty electricity filters too, but I do not have personal experience with them. If you have the means, I would suggest doing some research into these options and choose a whole-house system. Just be certain the system you choose will clean up the neutral and ground too.

Plumbing as Ground

In many towns and cities, instead of burying a grounding rod, electricians have attached grounding wires from homes to water lines. If there is dirty electricity on the ground in your home, or in a neighbor's home that is similarly grounded to their water pipe, the stray current can transmit down the water line and infiltrate throughout your home wherever there is plumbing. This is not a strong enough current to cause you to receive an electric shock, but it will irradiate VLF/LF of varying intensity.

To detect these anomalous currents, you may need to hire a building biologist to come into your home and survey the water line. These professionals have more sensitive equipment than the lower-priced handheld units available to the lay public. If you do have stray currents on your water line, you may want to consider remediation. Ask your plumber to cut a segment of the metal water pipe near where it enters the house in the basement, and have it replaced with PVC or another non-conductive material. This should be done in conjunction with an electrician to make sure both the plumber and electrician are comfortable with the arrangement and that the ground is not compromised in the process. In the usual scenario, cutting a short segment of house plumbing in this manner should not affect ground, as long as grounding was properly performed outside the house.

If you take these steps, your ELF-EMF and dirty electricity problems should be over.

CHAPTER 6

RADAR, MICROWAVES, AND RADIOFREQUENCY

We have been living within an environment infused with human-generated radio waves for over one hundred years. If it weren't safe, wouldn't we know by now? That is a hard question to answer. In general, safety depends on the amplitude, or signal intensity, and the frequency of the wave. Depending on a radio broadcasting antenna's location and its signal strength, it is likely people *have* been harmed, but an actual percentage is impossible to infer.

Some believe government has our back and properly regulates industry when necessary. Others are mistrustful with a healthy skepticism about the oversight process. Over the past few years, government shenanigans have become increasingly visible to the masses. It hasn't been pretty. We have all witnessed the revolving door between government and industry. The result is often protection for business leaders and their potentially harmful products at the expense of public interest. Add to that the power and influence of the military-industrial complex and potential threats to national security, and human health usually takes a back seat. A brief look into the history of microwave radiation and the development of safety standards is eye-opening.

The bandwidth between 100 kHz and 300 GHz within the electromagnetic spectrum has been classified as *radio frequency* (RF) radiation. This bandwidth is just above LF, the frequencies emitted by dirty electricity. *Microwaves*, or MW radiation, are a subset of higher RF frequencies spanning from 300 MHz to 300 GHz. Although the "micro" seems to imply that microwaves should have a wavelength that is a millionth of a meter, microwave wavelengths lie between 1 millimeter (a thousandth of a meter) and 1 meter. Microwaves are not produced in nature, at least not on Earth. But physicists and engineers have learned how to create them, leading to important applications, including delivery of therapeutic heat in healthcare, detection of an object's position through the use of radar, wireless transmission of information in the form of radio, and now wireless communication with cell phones, Wi-Fi routers, etc.

Medical Diathermy

During the early days of experimentation with MW radiation, the scientist von Zeyneck demonstrated in 1908 that RF radiation directed to the body could generate heat in targeted tissues, a technique he termed *medical diathermy*. Since the invention of the *magnetron* in 1920, we have been able to generate higher-energy, higher-frequency waves of radiation at increased amplitudes, providing the ability to increase the temperature of exposed tissues. Physicians explored medical diathermy as a medicinal tool for many years.

Diathermy often employed a frequency of 2.45 GHz, the same frequency used by a typical Wi-Fi router today. Early research by Krusen in 1947 revealed that, even though the skin could remain normal in temperature, *internal* temperatures increased with exposure to MW radiation. This was viewed at the time as a favorable effect when applying diathermy in physical medicine.[121] Microwaves were used to therapeutically heat body tissues by medical researchers into the 1950s because they were believed to be safe if used cautiously.[122]

RF radiation was even used at one time as a safe way to raise a person's core body temperature 2-3 degrees Fahrenheit. This medical practice was called *fever therapy* and was something previously only achievable by injecting a person with bacteria or biochemical toxins. Old reports documented that fever therapy was effective at combating disease. I found it interesting that one of the first review articles on this topic was published in the presently highly regarded journal *Radiology*.[123]

The idea of purposefully increasing your core body temperature by exposing yourself to microwaves seems like insanity today. It would be akin to putting someone in a microwave oven, albeit on a low setting! Yet this was an approved medical treatment. If you've ever warmed up food in a microwave oven for thirty seconds or so and then taken it out, you've no doubt noticed that some sections of the food will be hot, while the rest will be cold. Knowing what we now know, fever therapy as performed one hundred years ago was clearly dangerous. Heating from RF radiation is uneven and dependent on tissue characteristics, in particular, whether there are ions or water molecules in the tissue that will respond to EMFs. In order to get a measurable change in an organism's body temperature, localized areas of the body would have been at a much higher temperature than others. Higher temperatures cause proteins to denature and cells to become damaged.

This technique is comparable to several other radiological procedures performed at the beginning of the twentieth century that now seem barbaric. It wasn't that long ago, before computed tomography (CT scans), that if we needed to get a look inside a person's head, we performed a technique called *air cisternography*. This involved draining fluid out of the spinal canal, replacing it with air, and then placing the patient upright, so the air would rise into the skull. Filling the normally fluid-containing ventricles with air would outline the brain. After that torturous procedure, we could X-ray the skull and get a very limited image of the brain's contour to tell whether it looked normal.

Nonthermal Effects?

While experimenting with medical diathermy, physicians developed a slow but steady concern that there might be *nonthermal* effects occurring to tissues exposed to MW radiation. One of the early articles raising the question of nonthermal effects was written by Gossett et al. in 1924. In this paper, the authors found that exposing plants with tumors to microwaves caused the tumors to initially grow more rapidly than expected and the plant to die. The authors surmised that the effects could not be explained by heating alone.[124] Plant tumors are not malignant like they can be in animals. Plants can develop a tumor in response to an agent–such as fungi, insects, or bacteria–or "spontaneously," meaning it is not known why the tumor developed.[125] The tumors' change in behavior when exposed to MW radiation was an important early observation of nonthermal biological effects of RF radiation.

The controversy over the existence of nonthermal health effects from nonionizing radiation has existed ever since. Early research on short-wave energy, RF radiation, was performed to understand the potential negative health effects of exposing the body to this form of energy. An excellent historical summary of the research performed from the 1920s to 1960 was written by Cook et al.[126] Over those forty years, many research scientists performed experiments that documented nonthermal effects, but louder–more influential–voices, including some physicists, claimed that all the observed effects could be explained by tissue heating alone. Thus, those claiming the existence of nonthermal effects were put on the defense as early as 1929, when it was decided that "the burden of proof lies on those who claim any biologic action of high frequency currents other than heat production."[127]

Protocols were criticized, results questioned, and most scientists were ultimately convinced their conclusions had been incorrect and the only true bioeffect of RF radiation was heat. Looking in the rearview mirror, it is quite clear why there was such a heavy hand denying any potential nonthermal health effects from MW radiation. It was obvious early on that mastering this powerful technology would ensure global technological domination and accelerated unparalleled economic growth.

Radar

In the early 1930s, radio waves became an important tool used by the Navy and Air Force to detect unseen craft. Based on research performed in 1888, Hertz proved that radio waves could be reflected off metallic objects, just like light. This reflection of invisible energy–akin to an echo of sound–could be detected, timed, and monitored over time. *Radio detection and ranging*–commonly referred by its acronym *radar*–became quickly recognized globally as a useful military surveillance device. Early researchers in various countries experimented with different frequencies within the megahertz bandwidth. The Soviet Union chose 75 MHz. Early British radar was at 30 MHz. US scientists created radar systems for different applications, each using their own frequencies: 205 MHz for controlling anti-aircraft gunfire, 100 MHz for aircraft detection, and 200 MHz for navy ship surveillance. With the exception of some German radar systems, all early radar systems used frequencies below the microwave bandwidth.

In the late 1930s, technological advancements, including the invention of the *klystron* tube by researchers at Stanford and the creation of the *cavity magnetron oscillator* by British scientists, created the ability to produce higher frequencies of RF radiation from 600 to 95,000 MHz.[128] This led to the creation of "microwave radar," a more sensitive and specific technology than the earlier version. Massachusetts Institute of Technology developed hundreds of applications for MW radiation.

Nonthermal Health Effects from Radar?

Reports began to come out that governmental personnel exposed to radar technology were experiencing adverse effects, including temporary sterility, balding, and others. Military technical advisors believed that radar could not possibly be harmful because it was only a little higher powered than medical diathermy units. But the dismissal of health claims by military leaders was not satisfactory to concerned personnel.

In response, a twelve-month-long research study was performed on forty-five civilian personnel involved with experimental radar work. Yes, this was a very small sample size by today's standards. Physical exams, blood counts, and inquiries into pregnancy were studied in the cohort. The blood tests performed were rudimentary physical examinations, subjective and incomplete, and inquiries into pregnancy of questionable significance, especially as only half of the twenty married couples had children. Symptoms, such as headaches that disappeared shortly after exposure ceased, were dismissed as unimportant. The report, published in 1943 by Naval Lieutenant Commander L. Eugene Daily, confirmed the status quo. No health hazards were observed to be associated with exposure to radar and, furthermore, the RF energy of radar was "not different from that of other high-frequency radio or diathermy units of equivalent average power."[129, 130]

Personnel were not buying it. In an attempt to confirm the safety of microwave radar and improve morale, the Air Force conducted a second study, which took a more detailed look at the blood of 124 officers and enlisted men who had been exposed to radar. Laboratory results reported that red blood cell counts were lower in exposed personnel than the commonly accepted level, but these levels were felt to be "within the accepted range of normal and did not vary significantly from those obtained in the study of controls." The *reticulocyte* counts were "significantly higher" (1.7-3.3 *times* higher) in the exposed personnel as compared with the controls, but these too were also thought to be "within the normally accepted range." The authors concluded, erroneously, that "no evidence was discovered which might indicate stimulation or depression of the erythropoietic and leukopoietic systems of personnel exposed to emanations from standard radar sets over prolonged periods."[131]

A low red blood cell count means a diagnosis of anemia. An increased reticulocyte count in the blood indicates an increase in the body's release of immature red blood cells from the bone marrow, presumably the body's attempt to correct the anemia. This would be consistent with the fact that 16 percent of the study's subjects had high icteric indexes, indicating jaundice, which often results from excessive breakdown of red blood cells. The authors, however, dismiss this increase, despite finding "biliary tract pathology" in some of the subjects, as due to an outbreak of infectious mononucleosis. An outbreak that the twenty controls apparently successfully evaded. To conclude there was no evidence of an effect on the blood of personnel exposed to radar seems to be a gross mischaracterization and an attempted cover-up. But independent scientists were soon to uncover harmful effects could occur in animals exposed to microwave radiation. In 1948, researchers noticed cataracts developing in the eyes of animal subjects, in addition to testicular degeneration. At the time, MW radiation was not regulated . . . at all. Some researchers suggested that MW radiation should be treated with the same respect as X-rays.[132] Now, seventy-eight years later, I completely agree!

Doppler Radar

At about this time, a marriage between the physics of Doppler and microwave technology was taking place. Radar for many of us conjures up the image of a law enforcement officer on the highway holding up a radar gun directed at oncoming traffic. The ability to get information on motor vehicle speed from radar requires computations and an understanding of the *Doppler effect*. This concept was initially described by Christian Doppler, an Austrian physicist, in 1842. By using sound as an example, the principles behind the Doppler effect are more easily understood.

The Doppler Effect

Imagine you are on a long straight road and you hear the siren of an ambulance or police car coming toward you. As the vehicle approaches, the siren's pitch is high and going even higher if the vehicle's speed is increasing. But, as soon as the vehicle passes you, the pitch immediately drops. This change in pitch occurs because of a fundamental property of waves called the Doppler effect. Because sound waves travel at a constant speed from where they are produced, the waves become compressed if the source of the sound wave (in this case, the ambulance or police car) moves *toward* a listener, effectively increasing their frequency. The higher frequency produces a sound with higher pitch. If the source of sound moves *away* from a listener, the opposite is true. The distance between sound waves then increases, effectively lowering their frequency and reducing their pitch. This is an abstract concept that may take a few minutes to grasp. A visual may help: (Figure 6.1)

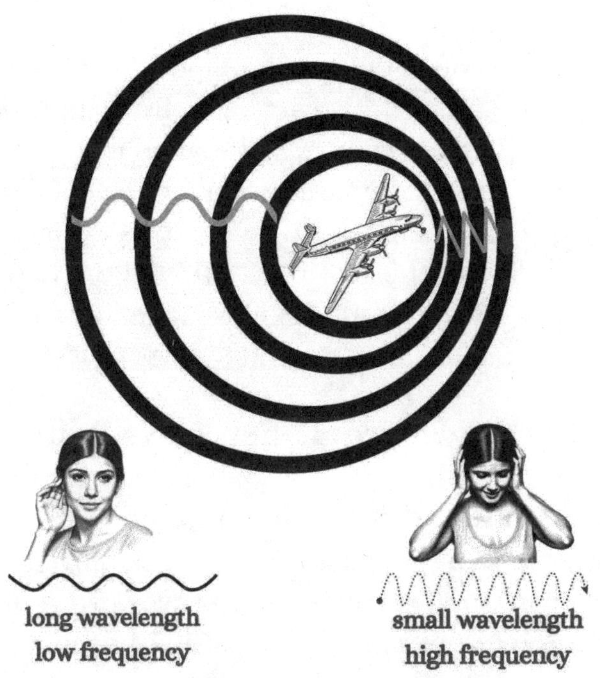

Figure 6.1. The Doppler effect

This is the same principle that allows an ultrasound machine with duplex Doppler software to determine the velocity of blood flow on medical imaging studies. With Doppler ultrasound, however, the ceramic crystal on the ultrasound probe produces sound waves and then listens for echoes while remaining stationary. It is the movement of blood cells that creates the Doppler shift. Software calculates the velocity of the flowing blood, which can reveal blood clots, heart valve defects, aneurysms, and other conditions associated with blood flow.

The Doppler effect occurs with electromagnetic waves too, but their velocity–"the speed of light"–is much faster than the speed of sound. A radar gun operates similarly to an ultrasound probe. The radar gun produces waves that it then listens for as the waves reflect off a distant object. As an object moves toward a radar device, the time it takes for the serial emissions of radiation to reflect back from its target to the detector decreases over time. The radar equipment is sensitive enough to pick up this tiny variation and calculate a velocity for the targeted object. The radar gun was designed by military researchers during World War II, and the first domestic deployment of a radar gun was in Connecticut in 1947. In the 1950s, this technology became widely used in radar systems.

Health Effects from Microwave Radiation

Hughes Aircraft and John McLaughlin

In the early 1950s, a significant cluster of employees at Hughes Aircraft Corporation working with microwave radiation suffered severe internal bleeding requiring hospitalization and blood transfusions. These employees were diagnosed with a rare condition called *purpura hemorrhagica* in which small blood vessels in the skin become damaged and leak. Individuals suffering with this condition present with localized areas of subcutaneous hemorrhage which look like purplish blotches, similar to those who develop a similar appearance from a minor injury when on too much blood thinner. About 75-100 cases of this condition occurred among approximately six thousand employees. The employees working with MW radiation all suffered headaches and some developed cataracts.[133]

John McLaughlin, a physician and consultant for Hughes, had been receiving disturbing data from other researchers around the country who were encountering disease in those exposed to MW radiation. McLaughlin wrote up these findings and produced a report in which he described purpura hemorrhagica, leukemia, cataracts, headaches, brain tumors, heart conditions, and jaundice as potential negative health effects from radar exposure.[134] After McLaughlin's report was disseminated, the medical community became wary of microwaves, and medical diathermy as a therapy fell out of favor. However, the surveillance ability offered by this technology continued to be of exceptional importance to the military-industrial complex.

Herman Schwan

A pioneer in biomedical engineering and biophysicist, Herman Schwan, who worked for the Navy and at the Moore School of Engineering at the University of Pennsylvania, believed that any effects associated with microwave exposure were only the result of tissue heating, so-called thermal effects. Given his credentials, Schwan was a very well-respected leader in the field. His premise was that calculations needed to be performed to assess the thermal absorption of various tissues based on radiation frequency and the exposed tissue's characteristics.[135] Schwan calculated that the body could dissipate a certain amount of heat generated from exposure to MW radiation over time and determined that any possible health effects from microwave exposure below 100 mw/cm^2 (the level of exposure known to cause heating) were insignificant. Based on thermal calculations, in 1953 Schwan suggested in a letter to the US Navy that a safe tolerance dosage of radiation would be 100 W/m^2.[136] Schwan was a revered international scientist, but it is important to understand that his original research was based on *theoretical* calculations. There were no ways to monitor actual dosage at that time; *dosimetry* wasn't available until after 1960.

Lockheed Aircraft and Charles Barron

In 1954, Lockheed Aircraft Corporation sponsored their own research to assess health effects on personnel exposed to MW radiation. This study, led by Charles Barron, was a similar study to the one performed by Lieutenant Commander Daily in 1943. Two-hundred twenty-six Lockheed employees and a control group of eighty-eight people were included in this study. Physical examinations, ophthalmologic exams, blood analysis, urinalysis, and chest X-rays were performed on each participant. Employees were also questioned as to the duration and manner of radar exposure and any appropriate marital and fertility history. As with prior studies, assessment of exposure was subjective and not confirmed with data from dosimeters.

A physician practicing in the twenty-first century might look at this list of exams under evaluation in the Lockheed study as somewhat primitive. But consider this was shortly after the discovery of DNA in 1953! Regardless, differences were observed in the health of those exposed to MW radiation when compared with control populations without exposure. Eye exams showed disease (aside from nearsightedness or farsightedness) in 12 of the 226 subjects (approximately 5 percent), presumably cataracts, as compared with only one case in the control group. Despite the glaringly increased incidence of eye disease in the exposed group, the researcher conducting this study did not confirm a connection between microwaves and cataract formation.

The Lockheed study also showed a significant decrease in the number of *polymorphonuclear* (PMN) cells, a type of white blood cell. Those exposed to MW radiation displayed 25 percent fewer PMNs as compared with those without exposure. A decrease in PMNs may indicate an impaired immune system. Increased monocytes and eosinophils were also found in the radar-exposed group. Elevation of these types of white blood cells is associated with an allergic response and is now commonly encountered in patients with allergies, but this wasn't known at that time. The researcher, Barron, could not explain the significance of these blood changes, and further studies did not provide any clarity. These abnormal results were eventually dismissed by Barron as an erroneous interpretation by the laboratory technician. The ultimate conclusion of this study was that there was no justification for public concern regarding microwave energy in the environment.[137]

Tri-Service Program

From 1957 to 1960, the government led a four-year research effort called "The Tri-Service Program," led by George Knauf, USAF. The program was not created to determine whether MW radiation was safe or dangerous. Instead, it was assembled to determine at what level thermal injuries could be expected to appear from exposure. For example, at which exposure level can cataracts be expected to form? At what exposure does testicular damage occur? Many academic centers in the US participated in this research. Some institutions studied pulsed microwaves, radiation emitted in short bursts of energy, while others studied continuous-wave radiation. Many frequencies were selected for study, including 200, 1,280 (1.28 GHz); 2,450 (2.45 GHz); 2,800 (2.8 GHz); 3,000 (3.0 GHz); 10,000 (10 GHz); and 24,500 MHz (24.5 GHz).

Most of the experiments were conducted at power densities over 100 mw/cm^2, the level at which thermal injuries were known to occur. Short exposure times were used; prolonged-duration and full-day exposures were not performed. Although some investigators raised concerns about nonthermal effects they observed, in the end the Tri-Service Program concluded that the effects of MW radiation were only due to heating and did not build up over time (noncumulative). It is unclear how the researchers concluded the effects were noncumulative, as no long-term studies were performed. The development of chronic health effects from long-term exposure could not be validly assessed.

Some effects described in microwave research at this time were bizarre. In 1960, an intriguing study by John Heller at the New England Institute for Medical Research reported that, at low levels, continuous EMFs caused unicellular organisms to "line up" and limited their ability to move. The direction of alignment varied depending on the frequency to which they were exposed. Not only were unicellular organisms shown to be spatially manipulated by EMF, the orientation of individual *intracellular* components was affected by microwaves, raising the question of whether or not MW radiation could have an effect on DNA.[138] When a cell divides, the DNA in the cell's nucleus needs to line up in a specific way to accurately reproduce itself and form two identical cells. If EMFs can affect the alignment of components within a cell, it is plausible that radiation might affect the cell's ability to accurately reproduce.

Based on his interpretation of the accumulated institutional research provided to him, Knauf recommended a maximum exposure level be set at 10 mw/cm^2, providing a safety buffer of a factor of 10 between it and the 100 mw/cm^2 known to cause tissue heating.[139] Ten mw/cm^2 is equivalent to a maximum exposure limit of 1,000 W/m^2, or 1,000,000,000 µW/m^2!

The Tri-Source Program ultimately decided there was no concern for human health effects from MW radiation. Effects were only thermal, not cumulative, and of little concern since "man has a built-in alarm system coupled with his threshold of pain that protects him from thermal injury."[140] It's hard to fathom how this was actually presented to the US Senate. Today, of course, we know this built-in alarm system does not exist. Furthermore, pain thresholds are unique to each individual. People with diabetes, for example, have reduced sensation, particularly in their extremities. Regardless of this inane conclusion, the Tri-Service Program report was instrumental in determining the safety standards for MW radiation exposure. An excellent review documenting the process that led to the important 1966 regulation can be found in the article "The Origins of U.S. Safety Standards for Microwave Radiation" in the journal *Science*.[141] Unfortunately, after the Tri-Service Program concluded, research into the health effects of MW radiation dropped off.

In order to conserve space, I included early studies in this discussion that I thought were particularly interesting. Zorach Glaser, PhD, put together an excellent compendium of over two thousand early research publications studying health effects from exposure to MW titled "Bibliography of Reported Biological Phenomena ('Effects') and Clinical Manifestations Attributed to Microwave and Radio-Frequency Radiation." This valuable governmental report cataloged over 120 health effects attributed to RF and microwave exposure, 40 of which were neuropsychiatric conditions.[142] This resource was approved for public release with unlimited distribution and may be found at ZoryGlaser.com.

The Fallout

Looking back, what can we take away from this early research? For one, microwave radiation can heat tissues, and the power density at which this is known to occur is at 100 mw/cm^2. During the early exploration of radar, people suffered physical damage from MW radiation that was not explained by thermal effects, such as purpura hemorrhagica, leukemia, cataracts, headaches, brain tumors, heart conditions, and jaundice, as outlined by McLaughlin. Lab tests showed effects on the immune system and anemia. But because no scientific mechanism had been defined at that time to explain these terrible clinical outcomes, this important information was dismissed as lab error. As is the case with other industries, whoever controls the narrative, controls the politics. Despite thousands of scientific studies documenting nonthermal negative health effects from MW radiation, the conclusion of the Tri-Service Program was that only thermal effects could occur. Knauf championed the opinion that eventually led to the "law of the land."

The Development of Safety Standards

Schwan's approach of performing calculations to determine radiation exposure led to today's standard of calculating *specific absorption rates* (SAR) to determine safety exposure limits of microwave-emitting devices. Understand that this approach assumes that the deterministic effect of tissue heating is the *only* potential effect from MW radiation and that nonthermal, stochastic effects *cannot* occur. This assumption is fundamentally flawed.

Setting regulations is not a cut-and-dried task. Data are typically disputed, and lobbying and corporate interests confound issues and influence lawmakers. In 1966, the US government set safety standards for occupational exposure to MW radiation. The limit adopted as the maximum safe exposure was set at 10 mW/cm^2 by Herman Schwan, chairman of the committee tasked with establishing this limit. This is cataloged as U.S. Standard C95.1-1966. Even though this limit was set for occupational exposure, unfortunately this standard affects all of us. The maximum limit for microwave exposure was set with insufficient scientific data and was influenced by industry and the military-industrial complex.

Microwave Ovens

During the 1970s, MW radiation found a new application. The knowledge that microwaves at a certain power density heat body tissues meant it might be useful for heating food, and microwave ovens came into existence. These devices produce a frequency of 2.45 GHz, the same frequency as a typical Wi-Fi router. The EMFs produced by the oven exert torque on water molecules and other molecules within the food, causing the molecules to bump against each other. This friction causes the food to heat up.

By federal regulation, microwave ovens must emit less than 5 milliwatts (mW)/cm² of MW radiation two inches from the oven surface. That equates to 50,000,000 µW/m²! This is a level of radiation five times more intense than the amount permitted for cell phones in the US and many other countries. Applying the inverse square law, manufacturers can produce microwave ovens that "leak" radiation and expose the user to the following intensities:

Distance from Oven	Intensity in µW/m²
2 inches	50,000,000
4 inches	25,000,000
8 inches	12,500,000
16 inches	6,250,000
32 inches	3,125,000
64 inches	1,562,500

At just over five feet from a microwave oven, it is therefore within the acceptable limits for your entire body to be exposed to an intensity of radiation that is just over 1,500,000 µW/m². Although I only used our microwave oven to reheat coffee, after surveying the radiation intensity produced by our appliance, I tossed it out about ten years ago. But I have surveyed many microwave ovens since, and they often do not meet this requirement. While on a recent trip, I surveyed the MW radiation intensity of a microwave oven in a cabin with no other sources of RF radiation. At a distance of six feet from the microwave, when the oven was turned on, I recorded maximum intensity readings over 2,500,000 µW/m², the detector's intensity limit.

Microwave ovens, radar applications, and medical diathermy all use non-modulated radiation, meaning the radiation is in the form of a smooth sine wave, like the 50/60 Hz waves in our electrical circuits. MW radiation, however, is at a much higher frequency. Because these frequencies are so much higher than sound frequencies, they can be easily *modulated* (by amplitude, *AM*, or frequency, *FM*) to carry sound and video information. This kind of modulation opened the possibility of wireless data transmission, leading to radio and television broadcasting and, eventually, wireless communication.

CHAPTER 7

AND THEN WE BROUGHT MODULATED MICROWAVES INTO OUR HOMES

Radiofrequency radiation is not only used to generate heat or detect objects and their speed. Early on, it was recognized that this form of energy could be used to transmit information. In fact, communication with radio waves (the bandwidth of EMR from 3 Hz to 1 GHz) preceded the development of radar by several decades. Through a process called modulation, information can be transmitted via EMR, as occurs with radio and television broadcasting and voice and data transmission. Modulation is an important concept to understand because it adds a dimension to microwave radiation that may cause additional effects on health.

Modulation

Musical instruments can provide a useful analogy to explain the concept of modulation. When we listen to the sound produced by an instrument, it is modulation that imparts character to each tone, something that distinguishes it from other sounds of equal amplitude and pitch. For example, imagine your eyes were closed and a note was played on a piano. You might not know what specific note you heard, but you would more than likely recognize the sound as coming from a piano. This is because the tone would have a specific characteristic, a quality that your brain would recognize. If a violin, flute, and French horn were all to play the same note at the same volume during a symphonic composition, your ear would be able to discriminate between the different sounds even though they were all coming at you with the same frequency, wavelength, and amplitude.

In music, the individual sound quality attributed to a specific instrument is referred to as *timbre*. Every instrument produces a sound with its own unique timbre. Even two violins playing the same note may be distinguishable because the timbre varies slightly between one violin and the other. Timbre is determined by a number of variables, including the material composing the instrument's body, the position of the holes in the instrument, the size of the instrument, the instrument's shape, and in the case of a stringed instrument, the string's composition. These variables each have an effect on the instrument's tonal quality. A nylon string will sound different from a steel string, which will sound different from a string made from bronze.

Timbre is created by the modulation of a sound wave. A tonal note is the product of a repetitive oscillating sound wave with a specific wavelength and frequency. Although sound waves are longitudinal waves, meaning they travel back and forth in the plane through which they propagate, they can be represented in a diagram as sine waves, just like EM waves. Changes can occur within the wave shape, but as long as the frequency remains the same–that is the distance between the peaks or the troughs remains the same– the note will retain the same pitch. Changes to the wave shape will, in a sense, provide additional information and create timbre. In Figure 7.1, the flute, voice, and violin are all producing the same note, as their wavelength is the same, but the wave shapes are markedly different among the three instruments.

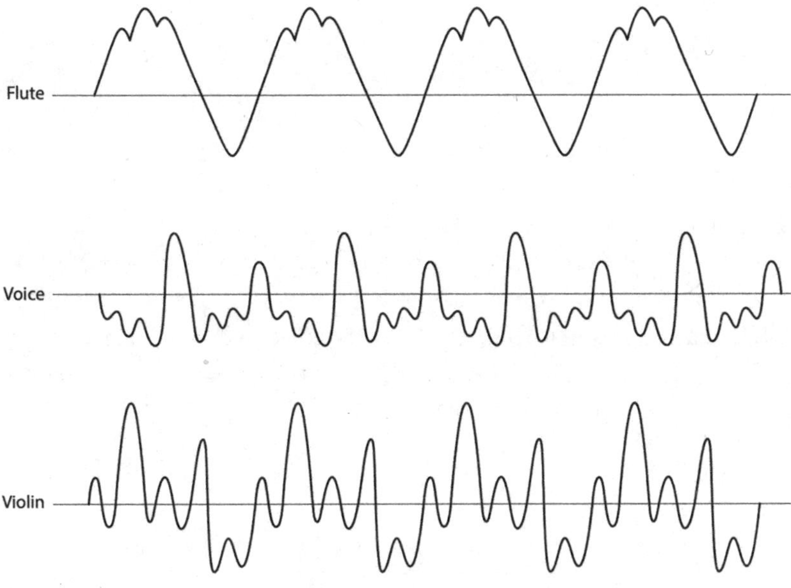

Figure 7.1. Timbre

Now, imagine you were listening to the following waveform:

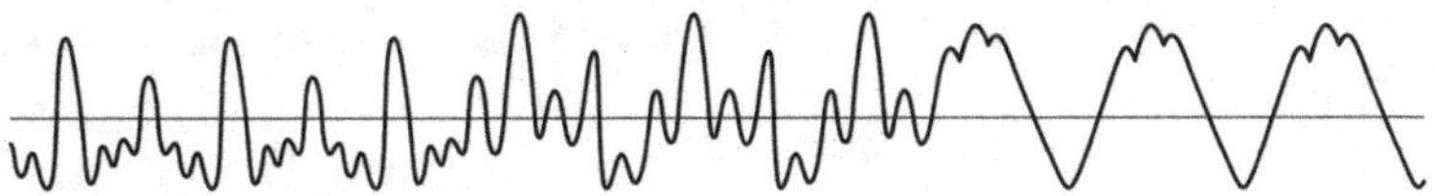

Figure 7.2.

You would recognize that first, you were hearing a voice, then a violin, and finally a flute, all playing the same note. The

frequency remains the same, but the modulation imparts additional information, i.e., what instrument is being played.

Now, imagine you were listening to several instruments
play the same note at the same time.

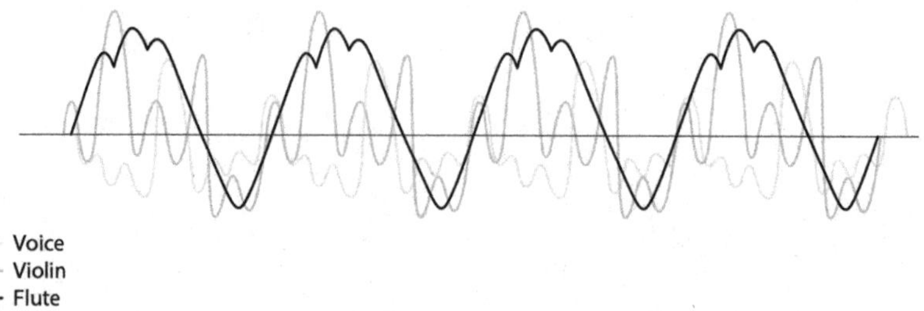

Figure 7.3.

It looks like a confusing jumble on paper, and in effect, the
resulting sound would be a single composite wave that looks
something like this:

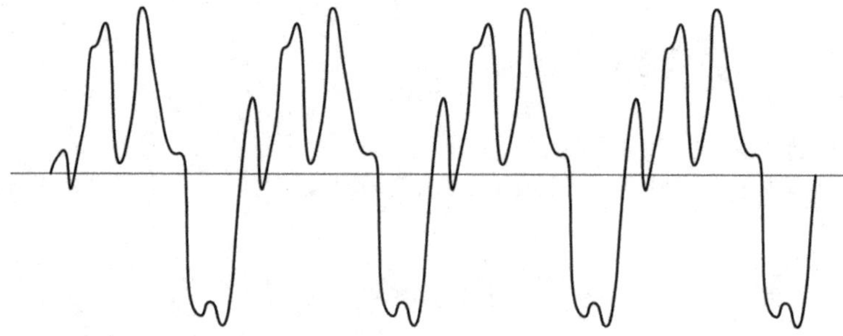

Figure 7.4.

The brain has the ability to separate out, or *demodulate*,
the individual sound waves. Most people hearing this mess
would be able to tell they were hearing a voice, a violin, and
a flute playing the same note at the same time. Conductors of
classical music orchestration make use of this phenomenon

by grouping or bringing out different "voices" through their interpretation of the works created by composers.

Engineers have long known how to modulate EMR, but their sophistication has increased dramatically over the past few decades. As a result, the complexity and volume of information transmittable by this technology have dramatically increased. In the beginning, modulation was achieved by varying a wave's amplitude and frequency, which led to the creation of radio and television broadcasting.

AM Radio

When most people hear the term radio frequency (RF) radiation, they think of radio. There are, of course, still conventional radio stations, but the trend for listening audiences has been to stream music and podcasts through wireless communication radiation (WCR) networks, which operate at a much higher frequency than traditional radio stations. Conventional radio and television broadcasts transmit within the lower-frequency bandwidth of RF radiation.

The invention of radio occurred in parallel with the development of the wireless telegraph. Both technologies were designed to provide distant communication by transmitting information via the modulation of radio waves. In 1901, the Italian inventor Guglielmo Marconi received the first successful transatlantic radio signal and is credited with inventing radio. During World War I, engineers figured out how to modulate a radio wave's amplitude to transmit encoded messages, a technology termed radiotelephony. *Amplitude modulation* (AM) of radio waves was then used to transmit audio signals that could be decoded by a receiving electronic device called an AM radio (Figure 7.5).

AM radio uses the bandwidth between 550 and 1600 kHz and was the main method for wireless broadcasting for many decades. Those of us who are older might remember driving in a car equipped with only an AM radio dial.

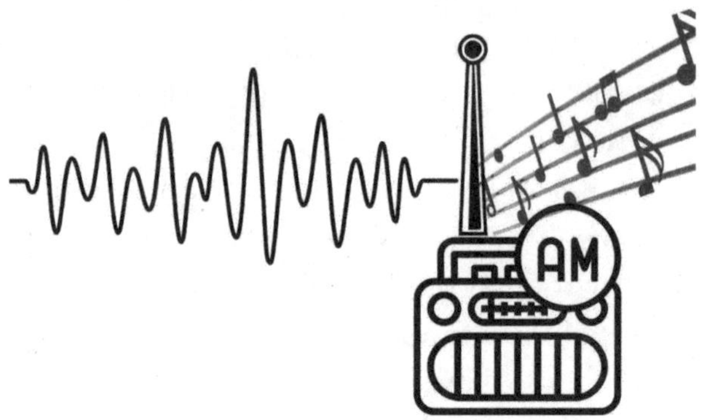

Figure 7.5. Amplitude modulation

FM Radio

Frequency modulation for radio broadcasting (FM radio) was invented in 1933 by Edwin Armstrong (Figure 7.6). By modulating *frequency* as opposed to *amplitude*, a high-fidelity sound transmission became possible. Because of the increased quality of the sound, it didn't take long for FM radio to become the choice for broadcasting music. Although the first car with FM radio was commercially available in 1952, it wasn't until the late 1960s that many cars (including ours) were equipped to receive these signals. A new frequency bandwidth, higher than that used by AM signals, was opened up for FM radio, mostly within the 87.5- to 108-MHz bandwidth, with few exceptions.

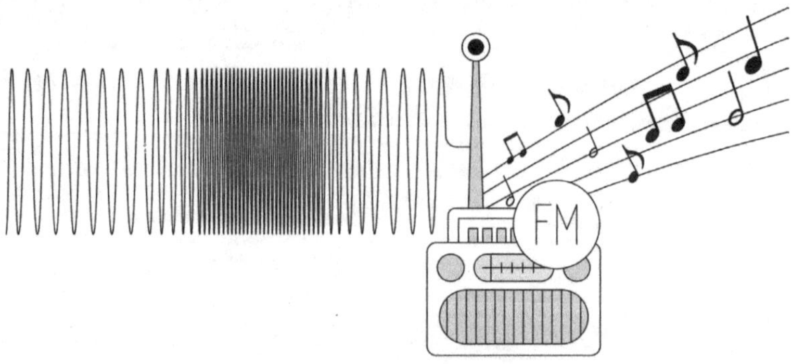

Figure 7.6. Frequency modulation

The low frequency of radio broadcasting radiation allowed their signal waves to travel far from their source and easily penetrate through most building materials and bodily tissues. The frequency bandwidths of AM and FM radio were, and still are, relatively narrow. In order to reach larger audiences, companies increased the power density, or strength, of transmission. Power densities for radio broadcast transmission could vary from several hundred watts to 45 kW. Antennas were often miles away from population centers.

Technological advancements enabling the transmission of video along with audio led to the creation of *television broadcasting*. The *television set* was first introduced to the American public at the 1939 World's Fair in New York City. Until that time, families often sat and listened to their radio for an evening's entertainment. By 1953, approximately one-half of US homes had television sets. Before streaming services, and cable before that, television broadcast frequencies in the megahertz range were emitted by the station's antennas. Television sets came with an attached antenna that received the station's signals. Television stations were given 6 MHz of bandwidth to transmit their signal. Good reception was crucial for an enjoyable experience. By the time the signal reached most neighborhoods from distant antennas, it had significantly weakened, but usually not so much that you couldn't enjoy the programming. It was extremely frustrating to try and watch a TV show if the station was so far away that you got electromagnetic interference, which showed up as "static," creating a snowy background that partially obscured the picture. Standard television broadcasting stations were licensed to emit frequencies between 54 and 890 MHz, with a power density near 1 megawatt. Back in the late '60s and early '70s, there were seven very-high-frequency (VHF) channels in the largest broadcast market in the country, New York City. Ultra-high-frequency (UHF) stations offered a few additional options.

Generating RF in the Home

Citizens band (CB) radios and amateur, "ham," radios were perhaps the first devices that brought the generation of modulated RF radiation into the home. Because radiation intensity diminishes with the square of the distance from its source, the radiation exposure a user would receive from ham radios, in particular, was much greater than that received from radio and television broadcasting antennas located miles away. Although communication on ham radio devices in the home was primarily a hobby (except in places too remote for the installation of telephones, like many areas of Alaska), CBs were an important method of communication between truckers and workers at outdoor job sites. These devices could be tuned in to one of forty channels within a frequency band between 26.965 and 27.405 MHz, centered at 27 MHz.

Digital Information and Computers

While some folks communicated via CB radios, computers in affluent high schools and universities were beginning to communicate with each other through hardwired networks. The internet was actually created in 1969, but it was in 1983 when the first *wide area network* (WAN) was created by the National Science Foundation (NSF) of the US. This network, called the NSFnet, provided users with access to a massive number of computers all physically interconnected. Rules, laws, and regulations collectively referred to as the internet protocol (IP) were created and later redefined to optimize resources. But this network was primarily limited to academic institutions.

Personal computers didn't become a thing until the late 1970s. One of my fondest memories of childhood was going to a computer fair in 1977 with my father and seeing this new company called Apple display a personal computer they called the Apple II. At the company's booth, a freestanding monitor displayed a never-ending loop of their logo dancing across the screen. I casually walked up to the keyboard and pressed the "control" and "C" buttons at the same time. The screen went blank except for a blinking green cursor. It was a keystroke I had learned on a computer terminal in middle school. My father was so impressed that I knew how to use a computer, he bought me an Apple II along with a color monitor!

It seems so silly to write now . . . the words *know how to use a computer* meant something very different then. Neither of us had a clue as to what lay on the horizon. Even back then, Apple computers were awesome. But they were just a hard drive in an attractive case. They were machines you could program in simplified languages like BASIC and PASCAL; there was no desktop or icons, and there was no "online" for home users. But the ability to program a monitor or display meant an enormous leap ahead from the series of numbers offered by programmable calculators, which had been the "go to" for complex calculations.

Internet

It wasn't until 1993-94 that the World Wide Web became available to nonacademic types and nonbusiness users. For most of us, life changed forever the first time we logged on to the internet and the World Wide Web. Early on, the internet was a novelty. Accessing the web required a connection between the computer and a phone line. When you wanted to access the internet, the computer could call an access number by using an external or, later, an internal modem, produce a series of tones, and then connect via an internet service provider. Typical connection speeds at that time were 40 to 50 kilibits per second (kbits/s), although rates could drop to 20 kbits/s in hotels and other "noisy" environments. Today's online content would not have been feasible with those slow connection speeds. Videos were not possible. Even a single picture could take minutes to download.

For many in the older generation, the internet was pretty much considered a toy. If one needed to contact an individual or a business, the white pages and yellow pages, respectively, a printed volume of addresses and phone numbers colloquially called *the phone book* that was distributed free of charge to telephone company customers, allowed your fingers to "do the walking." There were hardwired phones in most every home and public space. Everything was written down. If it had been typed into a computer, it was often then printed to ensure a hard copy existed. However, as technology brought a variety of improved connection options, including dedicated data lines, DSL connections, and T1 lines, the internet became reliable and easier to access. A dedicated (T1) transmission line from the internet service provider to a computer would allow data transmission at 64 kbits/s. Data compression software, coaxial and ethernet cabling, and a slew of other technologies brought hardwired connections into the gigabits/s range. External modems enabled the connection of multiple devices via ethernet cables to the hub, which was then hardwired to the internet service provider.

During the advancement of all this technology, there was no microwave radiation emission. Every device was hardwired.

Wireless Communication Radiation (WCR)

Advancements in modulation techniques enabled modulation of microwave radiation. Because the frequency of MW radiation was so much higher than the lower bandwidths of television and radio broadcasts, much more information could be transmitted in a given amount of time. The ability to transmit high volumes of data led to the proliferation of technology that dramatically changed the landscape of our homes.

Over time, everything that was once wired has become "wireless." Multicomponent devices requiring communication between different units brought wireless controllers for gaming consoles. Wireless keyboards, mice, and printers enabled communication between peripherals and computer hard drives. Remotes controlling TVs, VCRs, DVD players, garage door openers, and other appliances helped clear up the clutter of wired cable boxes and even eliminated the need to get up from the couch to turn devices on or off. These applications required limited radiation intensity, enough to span short distances, typically within the confines of a room so the signal could be picked up by the nearby receiver. Bluetooth wireless technology, introduced in 1998, provided and continues to provide short-range data transmission.

Bluetooth operates at 2.4 GHz, the same frequency as a microwave oven. But there are a few important differences between the two systems. First off, the power density emitted from Bluetooth devices is very low at 2.5 milliwatts, limiting their effective range. In addition, Bluetooth uses modulated signals, whereas microwave ovens do not. Rather than appearing as a smooth sine wave, the modulated waves of *wireless communication radiation* (WCR) are erratic, affecting the carrier wave's properties. Finally, WCR transmitted by Bluetooth is delivered in pulses, or bursts of frequency. This is unlike the continuous waves produced by microwave ovens.

By increasing the power density of emitters just a little bit more, inter-room communication devices became possible. Devices such as baby monitors, *digital enhanced cordless telecommunications* (DECT) portable phone systems, and wireless alarm systems became mainstream. The MW radiation produced by these devices was necessarily stronger so the radiation could transmit through walls and from one floor to another. Although some devices, such as baby monitors and alarm systems, only produce radiation when the systems are in use, other devices, such as the base station of DECT portable phone systems with multiple handsets, emit radiation constantly. Initially, all of these gadgets worked independently, meaning they didn't communicate with one another. But that changed with the development of a central hub, the Wi-Fi router.

Wi-Fi

Despite the fact that people sometimes think Wi-Fi and the internet are synonymous, the internet can still be accessed without Wi-Fi by hardwiring a device through an ethernet connection. But technologies build upon one another, and the development of wireless networks has certainly enhanced the internet experience. WaveLAN, the first wireless network, was developed by NCR in 1986 and became, in a sense, a prototype for Wi-Fi, the wireless standard ratified by the Institute of Electrical and Electronics Engineers (IEEE) in 1997. "Wi-Fi" is a trademarked phrase that refers to IEEE 802.11x standards. Wi-Fi is not an acronym for *wireless fidelity*, as many assume, based on the one used for high-fidelity sound systems (Hi-Fi). With the marriage of Wi-Fi and internet, users suddenly became exposed to much higher intensity levels of MW radiation and for much longer periods of time while surfing the web on their desktop computers and laptops.

In 1999, Wi-Fi was released to the public. Since then, it has gone through a series of embellishments, each enabling increasing speeds of data transmission and improved connectivity. The first Wi-Fi systems only modulated the 5-GHz frequency band and provided data transmission rates of 54 megabytes per second (Mbps). But starting in 2003, the 2.4-GHz frequency band was added. By emitting two different frequencies at the same time, routers enabled many more users and applications to be added without cross talk or interference. The higher 5-GHz frequency could transmit information faster, but the signal wouldn't travel as far. The lower 2.4-GHz frequency could more easily penetrate structures and travel longer distances, but couldn't transmit as much information in the same time frame. Adding a second frequency band allowed the incorporation of additional mobile devices and laptops coming to market, all designed to communicate wirelessly via Wi-Fi networks.

Over the years, data transmission rates have incrementally increased, with data transfer now in the multi-gigabit per second range. Wi-Fi 6E (extended) added a third frequency band at 6 GHz. If you have a Wi-Fi router, depending on its age, you are more than likely being continuously exposed to two or possibly three different modulated frequency bands of MW radiation when at home.

In addition to decluttering by enabling the removal of ethernet cords and cables, Wi-Fi routers allowed for a dramatic proliferation of devices that both send and receive WCR, enabling communication with the network and remote control from the router. Security systems and cameras, nanny cams, smart televisions, and many others came to market. Each device was a new source of potential RF emission. Smart homes and personal digital assistants like Alexa communicate through Bluetooth to appliances, lights, wireless speakers, and many others, offering a means to control devices through voice command. The number of appliances and gadgets that have been created taking advantage of this technology is mind-boggling.

Boosters or *extenders* have added yet another source of microwave radiation in the home. Although they both ultimately achieve the same goal, a Wi-Fi booster acquires a signal that has been produced by a router or access point, and rebroadcasts it, effectively creating a *second* network. A Wi-Fi extender extends the coverage area of your Wi-Fi network router by receiving the Wi-Fi signal, increasing its amplitude, and then transmitting the amplified signal. This increased intensity of Wi-Fi radiation often extends beyond the home's exterior.

We know this is true because if you close the door and windows to your home and turn on your tablet or phone and search for a Wi-Fi network, you will see anywhere from a few to dozens of potential networks to join. Your device is picking up the frequencies from each of these radiation sources in the nearby neighborhood. No doubt *your* Wi-Fi router's signal can be detected by your neighbor three doors down. It's fun to see the names of all the networks available, but the sad truth is, your body is absorbing radiation from all those sources. If we could see microwave radiation as we can the visible light spectrum, I believe most of us would be frightened by the extent to which our homes would glow from the numerous sources of energy.

Is This Safe?

This is the million-dollar question. We know that radiation can penetrate into and even through the human body, depending on its frequency and intensity. We also know that EMFs can affect calcium channels on cell membranes and result in increased intracellular calcium and oxidative stress. Determining whether or not a specific source of radiation is safe most likely depends on many factors, such as the radiation's frequency, intensity, modulation characteristics, polarity, and duration of exposure. In addition, whether or not the radiation is pulsed or a continuous wave emission is also an important variable. Perhaps equally important is the individual's health status and body composition.

Because of the scientific community's ongoing belief that the *only* potential health effects of MW radiation are thermal, safety testing only involves the power density of the radiation and the potential tissue heating from energy absorption. The specific absorption rate (SAR) is the only parameter regulators use to determine whether the power density of MW radiation produced by a device is too high.

$$SAR = \sigma \times E\,2/m_d$$

σ = conductivity of the material, E = electric field intensity, and m_d = mass density

Unfortunately, the technique for measuring this one parameter is flawed. According to Herman Schwan, the scientist to first suggest measuring SAR, the important factors determining MW radiation's effect on the body are *frequency* and *exposed tissue characteristics*. Conductivity depends on the specific tissue's water content in addition to other qualities. The degree of heating is therefore locally variable depending on the part of the body being studied. Even within a square centimeter of tissue, variations in density and molecular structure likely affect SAR. Anyone who has studied anatomy knows tissue variation is particularly evident in the head and neck. There is no way to create an accurate replica of the human body to precisely measure localized temperature changes from EMFs. Instead, industry uses *phantoms* to model and estimate temperature changes from MW exposure.

A commonly used phantom (Figure 7.7) is a hollowed-out head shape that can be filled with different oil-based or aqueous liquid substances. Temperature changes in the liquid material can easily be measured before and after exposure to a device emitting radiation.

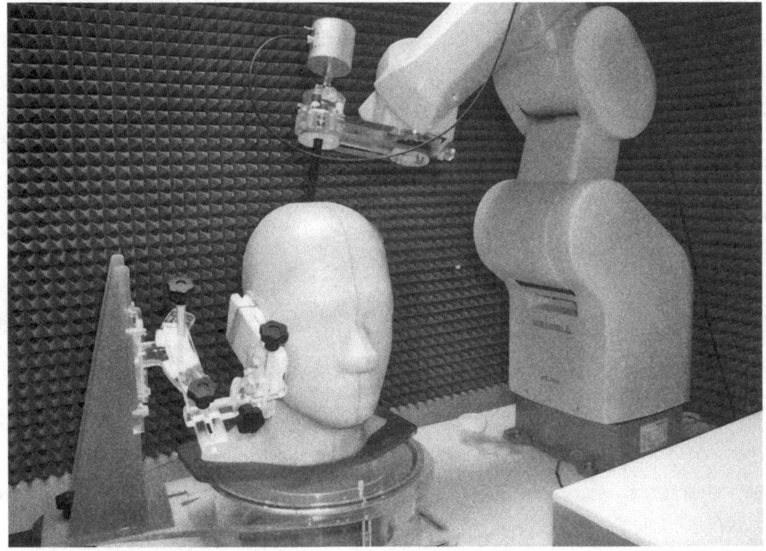

**Figure 7.7. Head phantom used to determine SAR.
Courtesy of the Environmental Health Trust**

This model provides a reproducible simulation for testing purposes, but does not take into account the complex anatomy of the human body. Beneath the skin, we are composed of many different tissues that vary in density and structure. There are bones, cartilage, fibrous tissues, areas of fat, and fluid collections. Structures have different lengths, shapes, and resonance characteristics. And let's not forget the metal ions locked in porphyrin rings or iodine accumulation in the thyroid gland. By observing X-rays, we know that some structures in the body absorb or deflect radiation, whereas others let it pass through more easily. The same phenomenon, more than likely, occurs with RF radiation.

Phantoms Can Provide Misleading Data

I understand the utility of phantoms, but it should be obvious they can provide misleading data. In the early part of my radiology career, I had an interesting experience with a phantom. I had developed an interest in bone and joint imaging early during my training, and one of my mentors, David Sartoris, a professor of radiology at UC San Diego, had given me a project to work on studying bone mineral density with a new technology called DEXA to help assess fracture risk. We knew that if one wanted to determine the risk for breaking a bone from a fall, it was necessary to measure the bone mineral density of *that specific bone*, because the mineralization of one part of the skeleton could be vastly different from another.

During my research project, I performed two reproducibility studies to test the accuracy of the technology. In one test, I performed three separate DEXA scans in one day on ten different patients. In the other test, I scanned ten patients, three times each, over the course of a month. My expectation was that if the machinery produced accurate data, the bone mineral density measurements would be similar to each other. I was optimistic because the manufacturer of the bone densitometry machine I was using claimed a 99 percent accuracy. Even though I scanned each patient myself, making sure to correctly position them, I could not get an accuracy anywhere near 99 percent. In some patients, the variability between studies was 10 percent! This was significant because it uncovered a limitation in the accuracy of the technology when used in a clinical setting.

I presented my findings at the annual meeting for the American Roentgen Ray Society (ARRS). During the conference, I browsed the bone densitometer vendor's booth and approached the engineer who designed the machine. I politely asked how he got such a high reproducibility. He proudly showed me a Lucite block with a section of lumbar spine encased within its interior. Yes, by aligning the block in the center of the mat and then pressing the button to scan the phantom, he was able to get results that were extremely accurate and reproducible. I was disturbed, but understood where the problem lay.

Unlike in their Lucite phantom, when a person lies down, there are subtle variations in the flexion or extension, tilting, and rotation of the spine and hips. These slight changes in position result in a difference in the overlap of bony structures. The X-rays produced by the bone densitometry machine therefore have a *different set of obstacles* in their path. This might mean that more X-rays are absorbed by bony structures or perhaps fewer will be. The bottom line is that a different number of X-rays will hit the detector on the other side of the body during a scan and yield a different bone mineral density reading. The technology is useful for determining one's category of bone mineral density but is less useful for following small variations in bone mineral density over time.

If the misleading data provided by phantoms is applied to a real-life clinical situation, it can result in inappropriate decision-making. But regardless, phantoms have been deemed an acceptable method for EMF safety testing. Government regulators accept this data from industry to "prove" their products are safe and in compliance with current exposure guidelines.

The Internet of Things (IoT)

Despite the continued concerns of many scientists and physicians on the potential negative health impacts of MW radiation, industry is marching toward a world in which everything is connected through wireless communication. MW radiation will emanate from all appliances; in addition to computers, phones, and lights, artificial Intelligence (AI) may orchestrate and perhaps ultimately control this international energy grid. This is called the *Internet of Things* (IoT).

Some may think this scenario seems really cool. To me, it is scary. I don't like the idea of enabling the utility company or anyone else to have the ability to hack into my home data and know what I set my thermostat to, how often I do laundry, or when I turn on and off my bedroom light. In addition, I believe it is important to move, to get up and turn lights or appliances on or off, and to have to think and remember. Although I live in a town in which autonomous driving vehicles are being tested, I prefer to have control of the wheel. In fact, I turn off the auto-assist feature in my car because I don't like having the car's computer taking over, even if I start to slowly veer into another lane. I like having the ability to make my own decisions, beneficial or not.

Regardless of my sentiments, the IoT is where things are currently headed. Be advised, this isn't just about a loss of autonomy or the capability for wide-scale surveillance. Never before have we been surrounded by so many sources of MW radiation in such close proximity as we currently have within our own homes. A phenomenon called *heterodyne* further complicates exposure levels, resulting in complex interference patterns in which there are additive and subtractive super-imposed frequencies emanating from each appliance. When I think about this, the hair on the back of my neck stiffens. The IoT will create environments inside and outside of the home in which our exposure to MW radiation will exponentially increase. As we build wireless infrastructure outside, it will become truly impossible to escape electropollution.

As if this weren't intrusive enough, implantable devices are being designed that upload personal health data to the inter-net and download information to deliver a tailored medication dose and/or alter the body's function. This technology, which is largely unregulated, is referred to as the *Internet of Bodies*.

CHAPTER 8

WIRELESS COMMUNICATION: NETWORKS, 5G, AND MOBILE PHONES

Whether to modernize your living space and bring technology into your home is a personal choice. There are people who prefer to live a more rudimentary lifestyle with some or none of these amenities. For the most part, it has been up to the heads of each household to decide how embroiled in technology, wireless or hardwired, they wanted their homes to become. With the exception of radio and television broadcasts, the outdoors used to be predominately electromagnetically clean (with the exception of radar used for weather assessment, highway patrol, airport, and military use). But in order to achieve mobile cell phone communication among users worldwide, networks were created. Networks consist of a multitude of antennas strategically placed throughout the environment so they can receive and transmit wireless signals from and to mobile phones. The installation of networks worldwide has dramatically altered the electromagnetic environment of the "outdoors," affecting not only the human population but—I suspect—all other life-forms as well.

Networks have been continually evolving for over forty years. They now provide the internet user with ultimate freedom. We are no longer confined to our homes if we want full access to the wilds of the internet. We can now stream video content and access other forms of data on mobile phones and other devices outside and far away from our home or office.

We are currently immersed in an experiment called 5G cellular network communication. 5G is an acronym for fifth generation. It does not mean the network transmits in the 5-GHz frequency band. The 5G media rollout has been extraordinary. For years now, it has been pretty much impossible to watch any form of media without being bombarded with an advertisement for 5G. Yes, trillions of dollars are being made by the world's governments selling off frequency bands to the telecommunications and other industries for each to have their slice of the pie. All of these vendors and service providers need a section of the electromagnetic spectrum to transmit their information.

In late 2018, 5G installation began and continued throughout the COVID pandemic. While the rest of us were sheltering in place, mobile telecommunication companies were busy putting up cell towers all over the world, preparing for a seamless blanket coverage of the 5G network. A repercussion of this network installation is the exposure of the human population to higher intensities of MW radiation in addition to even higher-frequency waves of radiation corresponding to *millimeter waves* (MMW). Because of these factors and other complexities, the 5G network has raised the eyebrows of scientists worldwide who are concerned about its safety. Despite alarm bells being sounded by some in the scientific community, Omdia, a global research organization that tracks technology development, reported that 433 million global 5G connections were added between the third quarter of 2021 and the third quarter of 2022, almost doubling the number of total connections during that time from 489 million to 922 million.[143] By the end of 2022, there were over 1.1 billion wireless connections, with industry forecasting 5.9 billion by the end of 2027.[144]

5G wireless technology differs from that of the previous networks, 2G, 3G, and 4G. Although many of us have lived through previous networks, each of which lasted approximately a decade, we may not have understood how they differ. In order to understand 5G, unravelling some details of previous generations of wireless communication networks will be helpful. An in-depth description of the physics and engineering behind each technology is way beyond the scope of this book and, frankly, my knowledge base. But this level of detail is not necessary to grasp concepts. My goal is to convey, in an understandable way, how wireless communication has affected societal interactions in addition to our physical health.

The Genesis of Mobile Phones

Bell Telephone introduced the mobile phone in 1946. In fact, the first mobile phone call was made from St. Louis on June 17, 1946. The "car phone," created by Bell Laboratories, allowed users to place and receive calls from their automobiles. By 1948, there were five thousand subscribers to this service. Back then, mobile phone service was only available in a few major cities and along certain highway corridors, enabling communication while traveling in a moving vehicle.

In 1965, AT&T improved the mobile telephone experience by creating the Improved Mobile Telephone Service (IMTS). Although the system allowed many callers to place calls at the same time, the system was set up to service forty thousand customers, two thousand of whom lived in New York City. The wait time to place a call was said to often be thirty minutes or longer. It's hard to fathom that wireless technology existed so long ago. But the service wasn't widespread, and devices certainly weren't portable. The phone hardware filled the car's trunk, and calls were made through a switchboard operator.

In 1973, the first handheld mobile phone was used by Martin Cooper, an executive at Motorola, to place a call from New York to his company's competitor, Bell Laboratories in New Jersey. His team had created the world's first phone that could be held in one's hand. It was shaped like a brick and weighed over two pounds and therefore couldn't fit in one's pocket, but it was an engineering marvel that literally changed the world.

Nineteen seventy-nine proved to be a pivotal year during the early development of wireless communication. That was when the Japanese utility company Nippon Telegraph and Telephone (NTT) launched the first commercial cellular network. It was also the year *satellite phones* were first developed. The satellite network enabled users to communicate with each other by connecting with a satellite system orbiting Earth, as opposed to a network of terrestrial, or ground, antennas. As a result, satellite phones provided the ability to communicate while on the ocean or in regions devoid of antennas. If you have ever used a cell phone in remote parts on a trek, you or someone in your group were likely equipped with a satellite phone. Satellite communication networks have evolved like all others and offer voice communication in addition to SMS text messaging, email, and other smartphone functions.

The Networks

The first generation of *wireless cellular technology*, called 1G, was introduced in the early 1980s. 1G was an analog system made possible by the installation of network "cells." A cell is the name given to an area able to receive wireless transmissions from a centrally located tower or base station within a cellular network. The cell's radius is dependent on the frequency and amplitude or power density of the radiation emitted from the tower. The power density can be raised or lowered at the discretion of the network provider.

The first mobile networks were deployed in Japan and Scandinavia. In 1983, 1G came to the US when the first commercial cell towers were erected in Chicago. 1G primarily transmitted in the MHz frequencies. Because 1G frequencies were relatively low, they could travel long distances. Therefore, like radio and television broadcasting towers, antennas could be placed far from population centers and service a large area. Speeds were slow at 2.4 kbps, but being an analog transmission, 1G offered voice calls that reproduced sound more accurately, providing a more realistic, comforting sound than the digital networks that followed. By 1991, over 50 percent of the population in Finland and Sweden were utilizing the 1G network, and approximately 21 percent of the US was on the 1G cellular network.[145]

On the experiential side, the first pocket-sized cell phones were flip phones and were wonderfully convenient. No longer did we have to wait on a line to use a pay phone at a street corner. Needing a dime or a quarter to use a pay phone at the airport or in a restaurant was no longer an issue. You could carry a phone with you and put it in your pocket (something I no longer do)! It was life changing.

In 1991, *second-generation cellular networks* were introduced, called *2G* for short. There were competing systems; the US company Qualcomm developed the CDMA (Code Division Multiple Access) standard, while the Europeans developed the GSM (Global System for Mobile Communications) standard. The CDMA network opened up a wider frequency bandwidth and therefore enabled more users to access the service. Once launched, the network generated an explosion of users. This network put an end to analog voice calls. It was a subtle effect, but one that removed the warmth and complex timbre of a person's voice on the receiving end of the call. The familiar voices of those you loved and resonated with on the analog system were suddenly digitized, creating a subtle but real sense of disconnect. In exchange for this loss, the 2G network converted calls to digital information, which allowed for greater efficiency. Importantly, data and voice could now be encrypted, making it much more difficult for people to listen in on conversations.

Digitization enabled data transmission through the 2G network, including a new form of *short message service* (SMS) to send a display of words to a recipient. Until this time, *texting* was not a thing. Pictures converted into digitized format and multimedia could also be transmitted over the 2G network. But if you recall those days, you'll remember that transmitting a single image could take minutes. As texting slowly became a preferred method for communication, it dramatically changed culture, sacrificing etiquette and an audible form of connection to a more immediate but less personal one. The ability to transfer data wirelessly also led to the ability to download software onto a mobile device, the innovation that led to today's *smartphones*. In 1999, access to the internet became possible through the 2G network in Japan. *Personal digital assistants* (PDAs), like the BlackBerry and the PalmPilot, became widely popular. By the mid-2010s, 2G was used by over 3.9 billion people worldwide.[146]

When 2G first came out, the GSM network was limited to the 900-MHz frequency band, similar to 1G. Each channel was divided up into 200 kHz. This was a slow network with a maximum transfer speed of 40 kbit/s. Over time, frequency bands used by the GSM network increased to include 900- and 1800-MHz bands in Europe, Asia, Africa, the Middle East, and Australia, whereas 850- and 1900-MHz bands were utilized mainly in North and South America. For the most part, frequencies used by the 2G network were in the same range as those used by 1G.

Interpreting frequency bands can be confusing. In general, when a frequency band is assigned to a network, say 900 MHz, that number represents a range of frequencies within a bandwidth. Industry divides up the bandwidth into sub-bandwidths, one to send information out to a device, and the other to receive information. For example, in 2G, the 1800-MHz band represents a bandwidth from 1710 to 1880 MHz. The uplink ranges from 1710 to 1785 MHz and the downlink ranges from 1805 to 1880 MHz. Knowing the numbers of specific frequencies is less important than understanding that the network exposes the environment to a spectrum of RF radiation, not a single frequency.

An Out-of-This-World Coincidence? Maybe Not . . .

With the installation of network antennas worldwide, global communication became feasible and, in 1994, the World Wide Web as we know it was born. In its wake, the electromagnetic environment of the Earth began to change, particularly in population centers. Most people were oblivious to this electropollution, but perhaps others weren't.

Have you heard of the Ariel phenomenon? This well-documented encounter with alien life-forms occurred in September 1994 in Rua, Zimbabwe, and in 2022 was described in detail in a film called *Ariel Phenomenon*. A large group of sixty-two grammar schoolchildren reportedly witnessed a UFO land on a hill behind their school while they were out playing in the schoolyard. Video testimonials and interviews were taken of the children as they explained what they saw and what they learned from the extraterrestrials. A professor of psychiatry from Harvard University, John Mack, MD, interviewed the children and recorded their observations. The story has been reenacted and events recounted in several recent documentaries. The children described the alien beings as being small, "one meter high." They were wearing silver space suits and had large black eyes. When asked what they had learned from the aliens, the children stated they had received telepathic messages that "technology was bad." That technology was damaging the Earth.

I personally found the stories and the interviews with the psychologist compelling and took a moment to do an online search to see what technology had developed in 1994. The first item listed on Google was "Birth of the World Wide Web." I thought it was a curious coincidence. But in a way, it also made sense. Aliens from another planet or craft wouldn't be concerned about our ability to create genetically modified organisms (GMOs) or how badly we contaminated our air or water pollution levels. How could they know? But electropollution, the human generation of modulated MW radiation, would be obvious to an advanced civilization. All of the frequencies emitted by cell phones, Wi-Fi routers, and cellular network antennas could theoretically be detected by intelligent beings beyond our planet. All EMR travels at the speed of light.

Although most of the documentaries on this encounter have concentrated on whether or not it happened, after listening to the testimonies of the schoolchildren and Dr. Mack, I was more concerned as to why these beings would tell a group of schoolchildren in sub-Saharan East Africa that technology was bad. These children had no experience with technology in other parts of the world. This community didn't even have reliable electricity. Perhaps as representatives of the human species, they were more easily approachable and potentially malleable than a group of schoolchildren playing in, say, Silicon Valley. Were the "visitors" going back to "Eden" to warn about eating fruit from the metaphorical tree of knowledge? Is uncontrolled electropollution with modulated energy waves disturbing the foundational energy grid orchestrating life on Earth? What were they trying to tell us?

Unfortunately, for all intents and purposes, the Rua encounter got mired in the fantastic debate as to whether or not the children were credible and whether there was any real evidence. The visitors' message did not have any effect on the development, production, or utilization of wireless communication technology among humans.

3G

Meanwhile, in the 2000s, 3G came to market and was the main network in operation between 2001 and 2010. Because 2G was becoming increasingly widespread and the usage and demand for portable data transfer was growing, the network needed to accommodate a greater demand. Technological advancements enabled faster data transfer rates, which rose dramatically in the mid-2000s. This allowed the introduction of smartphones and video calls. It utilized the UMTS (Universal Mobile Telecommunications System) standard, based on the GSM standard. Multiple bands of RF were opened up and used by the 3G network, including 850, 900, 1700, 1900, and 2100 MHz, a higher-frequency band of MW radiation than those used in the 2G network.

4G

LTE (*Long-Term Evolution*) was introduced to the Scandinavian market in 2009. This standard allowed for even faster internet access outside of the home. Offering speeds up to 1 Gbps, 4G LTE is still in use today. It utilizes numerous frequency bands from 450 MHz up through 6 GHz, an even greater number of frequencies in the microwave frequency range than its 3G predecessor. With 4G, the microwave frequency bandwidth from 2.1 to 6 GHz became available for networks to license.

The reason so many frequency bands are required for 4G to operate is, in part, because of the number of users. But it is also because of advanced technology used by this system. 4G LTE antennas use *Multiple-Input, Multiple-Output* (MIMO) technology, which significantly increased data speed and improved signal quality. *Multiple input* means there are anywhere from two to eight antennas beaming signals to your device (Figure 8.1).

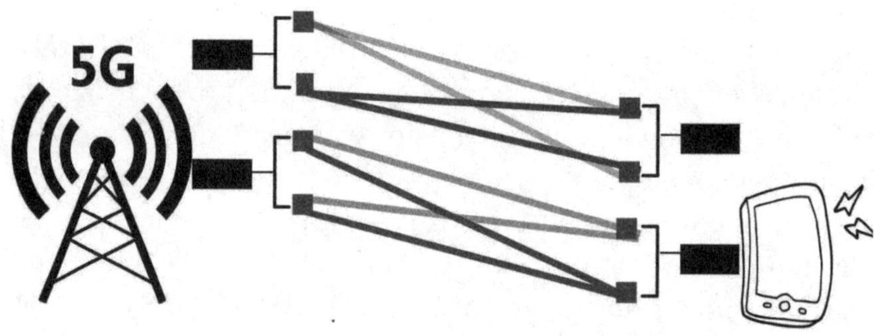

Figure 8.1. Multiple-Input, Multiple-Output (MIMO). With MIMO technology, multiple antennas are placed on both the cell tower and on the device, allowing for multiple potential pathways of communication.

The multiple streams of radiation improve the signal quality and network performance. *Multiple-output* means there are multiple antennas on the device accessing the network. Because the different antennas are spaced apart from each other, this arrangement provides different pathways for the signal to ultimately reach its target. If there is an obstacle in the way of one of the energy beams, another path may be free and clear. The result is a more reliable service to our devices. Another characteristic of the 4G LTE MIMO network is having multiple streams of data being emitted from different antennas to your device at the same time. This is called *spatial multiplexing*. This is a great method for getting larger amounts of information to your device in a shorter amount of time. With MIMO technology, the ability to focus the beam of radiation and direct it toward its intended recipient, a technology called *beam forming*, is possible, allowing for even faster, targeted delivery of data. MIMO technology offers higher bit rates of information to our devices, but it achieves this by exposing us to many more concurrent streams of RF and MW radiation. MIMO technology has dramatically increased our RF radiation exposure, as compared to older networks.

5G

I, like most people, thought I was on the 5G network when my cell phone suddenly displayed 5G on it. I was mistaken because the 5G on the phone meant I was using the 5-GHz frequency band. The term *5G* encompasses an enormous bandwidth of potential frequencies for license that include the 4G LTE frequencies up through 95 GHz. Industry has divided this network into low-, medium-, and high-frequency bandwidths. Many internet providers are closing down the 3G network and repurposing the bandwidth for 4G and 5G networks.

The only way I can think of how to impress upon you how wide the 5G bandwidth is to refer to an old-style AM radio. If you remember, to select a station with an old-style AM radio, you would turn a dial that would change the frequency of the radio's reception. The dial spanned a range of frequency from 540 to 1700 kHz. Each station was given a bandwidth of 10 kHz. Therefore, if you do the calculation (1700 - 540)/10, there were 116 potential stations within that AM radio band-width. One would need to turn the dial perhaps six times to scroll through all those stations in a car radio. To compare that to the 5G network, using a conservative window of frequency spanning from a low of 600 MHz to 54 GHz, if we compare apples to apples and consider the same bandwidth of 10 kHz per station, the number of potential stations would be (54,000,000,000 - 600,000,000)/10, or 5,340,000,000 potential stations. In a practical sense, if you were to access all of these stations with a knob as you did for the AM radio, you would have to turn that knob about 276,254,527 times to get from one end of the 5G spectrum to the other. That is truly mind-boggling.

The bandwidths of the 5G network are not the same as for AM radio stations. This is merely an illustration to help you visualize the enormity of the 5G bandwidth. What is even more sobering is that these are the frequencies now currently being used by wireless carriers. Licensing is available far higher into the millimeter wave (MMW) bandwidth, up to 95 GHz! It has opened up an enormous spectrum of RF radiation that is more expansive than AM radio, FM radio, TV broadcasting, 2G, 3G, and 4G combined. In addition, there are several features of the 5G network that make it different from previous generations of wireless communication networks.

Why Are There So Many Cell Phone Towers and So Many Antennas on Each Tower?

We have all been witnessing the proliferation of cell towers popping up on cityscapes and in rural landscapes. One day, we notice a new tower and we can't remember if it was there yesterday.

In addition, you have no doubt noticed an explosion of small cell installations on top of utility poles and buildings over the past few years. During the installation of 5G infrastructure, 4G LTE antennas were moved from outskirts into towns and cities along with the 5G antennas (Figure 8.2).

Figure 8.2. 5G antenna installation

Much to my dismay, one of these small cells was installed on a utility pole directly across the street from my house during the COVID pandemic. As it was being installed, I watched and filmed as the servicemen affixed three antennas on the top of the pole, each oriented at 120-degree angles to one another, providing a 360-degree area of coverage. After the antennas were installed, the men covered the apparatus with a light brown, semi-gloss cover to "beautify" the hardware. It looks like a dirty Q-tip. Most people walking or driving by the antenna have no idea what it is; some don't even notice it until it is pointed out to them. Once you know what they are, you see them everywhere. It can be frightening to suddenly become aware of something that has become so invasive in our neighborhoods and towns and yet not know when it was put up or what it is.

Why are there so many antennas, and why are they now in our towns and cities instead of only in far-off industrial parks and mountaintops? It has to do with physics. In general, with increasing frequencies of radiation, through the process of modulation, greater amounts of information can be transmitted more quickly. This is because more waves are generated and transmitted per second. But higher frequencies do not travel as *far* as the lower frequencies and are more easily blocked and absorbed. Because of this, cell towers emitting these frequencies need to be closer to the end user, which means cell phone towers need to be placed within population centers. In fact, MMW-producing antennas are needed every couple hundred feet to receive uninterrupted cellular service while in transit. Service is provided by one small cell, and then the signal is handed off to the next, and then the next. In order to get the frequencies into the device and past obstacles, such as leaves and rain, the intensity of the radiation emitted by these cells has increased.

The proliferation of cell phone towers providing the 3G network went, for the most part, unnoticed. Most people were happy for the increased prevalence of cell towers because it meant fewer "dropped" calls, which were incredibly annoying. Many of us shared the familiar experience, calling out, "Are you there? Can you hear me? Can you hear me now?" Sometimes just walking to the other side of a room was enough to lose a call or get clearer reception to continue a conversation.

In some communities, the erection of cell towers has caused alarm. Like many other public utilities and services, most people don't want the infrastructure to affect their view. "Not in my backyard!" Concerns were mainly regarding a drop in property value and the community's aesthetics rather than a drop in the community's health. In some demanding municipalities, cell towers were converted into pieces of terrible "art," including mechanical-looking pine trees, desert cacti, flagpoles, and bell towers, a process called *stealthing*. They all looked quite ridiculous. But an interesting thing about human perception is that over time, even though these obtrusive structures dot the landscape and cover our buildings, water towers, and church steeples, most people don't notice them after they've been up for a while. But regardless of whether or not you notice them with your senses, your body is being irradiated by these antennas 24/7. I performed a survey of thirty-five "Main Streets" in Pennsylvania during 2021 and found that those cities with 4G LTE antennas in town were exposing their population to anywhere from ten to one hundred times the radiation levels as the towns without these antennas. At street level, the pole-mounted antennas were associated with the highest intensities of radiation when compared to those mounted on building tops and towers.[147]

When companies put up cell towers in areas that don't have coverage, that is referred to as *growth*. A company's decision to place a cell tower in a given location is based on growth, capacity, and reception, as well as other parameters. Capacity increases with the installation of each cell tower as more calls can be handled within the cell. If there are an increasing number of subscribers in a given area and they are using the network for more minutes per day, the number of cell towers in an area will need to be increased. Each cell tower can only handle a certain number of calls based on the number of separate frequencies it can produce. If a large number of users are using separate frequencies within a cell at the same time, calls can be diverted to another cell. The power density of the transmissions emitted by towers in close proximity to one another is reduced to eliminate interference with neighboring cells using the same frequency. The seemingly ever-increasing number of antennas is therefore a reflection of increasing demand, as well as increases in competing vendors and service providers.

Millimeter Waves Only for an "On-demand" Service?

Network providers claim the high-frequency millimeter-wave bands are reserved for video streaming and other applications requiring high data throughput. If you want to watch a video on a 5G-compatible cell phone, when you start the video, your device will call up the 5G network on the local small cell. The small cell will then blast a prolonged pulse of MMW modulated with the desired content directed at your device and turn off again after the information has been received. Sounds good, but while surveying for 5G millimeter wave frequencies, I've noticed these antennas constantly beam MMW radiation at significant intensities, even when they are not being engaged. What happens when you put a body in between the cell phone and the transmitting antenna? Does the radiation penetrate through the body to reach the phone, or does the movie stop buffering and disconnect from the network? Given that video streaming is getting smoother and smoother, we can assume at least some MMW radiation penetrates the body. What is the health impact? No one knows, but there is no reason whatsoever to assume that it is minimal.

Furthermore, the resonant frequency for water vapor is 24 GHz, and for oxygen is 60 GHz. As previously mentioned, the resonant frequency refers to that frequency at which maximal energy absorption occurs. Twenty-four GHz and 60 GHz are both within the 5G bandwidth. Does the energy absorption of MMW by water vapor molecules and/or oxygen have a biological effect when we breathe them in? Does energized oxygen bind with hemoglobin? If so, is there any biological effect? Importantly, is there an environmental effect? A *heating* effect of the air itself? These are some of the questions I hope have been answered or at least considered before installing this technology globally.

Massive Data Transmission

The MIMO antenna technology used in the 4G LTE network was significantly boosted for the 5G network. The technology has been renamed and is now called Massive Multiple-Input, Multiple-Output (Massive MIMO). Instead of a maximum of eight antennas present in the LTE network, with Massive MIMO, a grid of anywhere from dozens to hundreds of antenna units are placed on each antenna panel, allowing for multiple antennas to focus EMR on multiple individual users at the same time. For example, a 64×64 downlink configuration would mean that there are sixty-four antennas on the cell tower beaming RF radiation to sixty-four different cell phone antennas receiving information. Like standard MIMO, multiple antennas per user allow for *multiplexing* data, that is, sending multiple streams of data at the same time, as well as duplicating data, increasing the quality of signal transmission. Because there are so many antennas producing RF radiation, multiple devices and users can be serviced at once. If you look up at a collection of antennas on the roof of a building, you will see a series of oblong (4G LTE) antennas and a smaller square antenna. The square antenna is the 5G antenna (Figure 8.3).

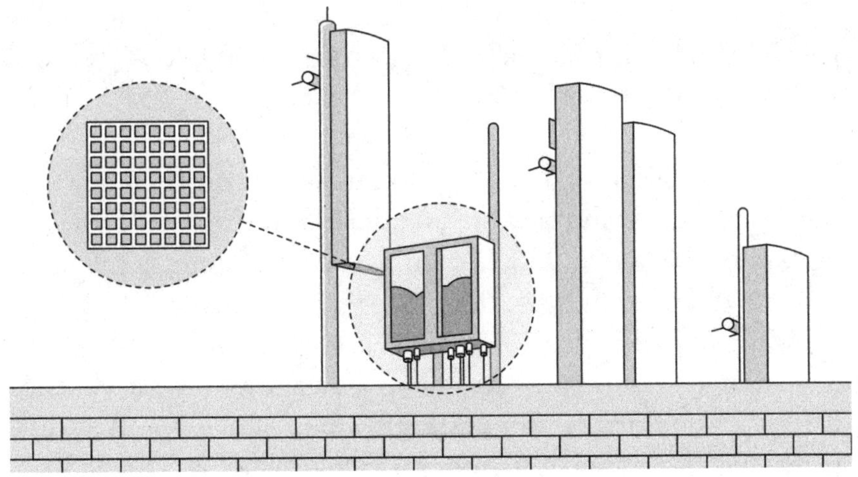

Figure 8.3. The 5G antenna

Beamforming

Massive MIMO allows for three-dimensional beamforming, reducing the exposure for those who are not using the high-frequency band. But the general public and the rest of the environment are continually irradiated with the low- and middle-frequency bands of the 5G network, just as with 4G LTE. Whereas the low and middle frequency band antennas beam out in 360 degrees, creating electropollution and exposing everyone and everything to radiation in its path, beamforming of the high-frequency band focuses the transmission to the device that is asking to download data. Focused radio waves are stronger and transmit longer distances with more intensity, so they can more easily penetrate through structures, such as buildings (and people?). This is akin to focusing sunlight with a magnifying glass. But if you've ever focused sunlight through a lens onto a leaf, you've probably noticed that if you keep it there too long the leaf gets singed and eventually catches fire.

That's where our wireless networks are at the moment. Despite the ominous presence of antennas throughout the environment and the ubiquitous electropollution these networks have created, the most intense exposure to MW radiation most of us receive is from a device sitting in our pockets.

Cell Phones

Carrying a device that enables you to access pretty much anything and everything is mind-boggling for those of us who grew up before this technology. When the Apple iPhone came out in 2007, it forever changed the cell phone market. Smartphones were a game changer. Personally, it took me a while to board the smartphone train. I was comfortable with a flip phone and–like so many–resistant to and suspicious of change. But, after truly considering what the iPhone had to offer, it seemed like a no-brainer. My pockets were already stuffed with an iPod onto which I had uploaded my old CD collection; a PDA that kept track of my calendar, address book, and to-do list; and a flip phone. Carrying three devices in my pocket was cumbersome. When I realized I could do away with all three and replace them with one phone, I went for it. Having all my information on a device that could easily break if I dropped it or accidentally got it wet seemed a bit risky, but the convenience and compactness of the technology were unbeatable.

Cell Phone Antennas

Cell phones transmit RF radiation to and receive radiation back from nearby cell towers. The quality of reception is based on the distance between the two, the signal intensity, and the frequency being transmitted. Typical cellphones have enough power to reach a cell tower somewhere between twenty-two and forty-five miles away. However, obstacles blocking the line of sight between the tower and phone, including trees, buildings, hills and valleys, and any other structures, can reduce transmission quality. A longer distance to the cell tower and a greater number of obstacles requires cell phones to produce stronger, more intense signals. That's why cell phones should not be used in areas with poor reception, such as inside an elevator.

Older flip phones produce radiation only when they are in use. But smartphones nearly continuously emit MW radiation regardless of whether or not you are actively using the phone. Apps, including email servers, messaging services, etc., constantly ping the network to fetch for updates. Because apps send and receive information even when not actively engaged, they dramatically increased the need for more wireless communication. Lower-frequency bands became saturated as the number of cell phone users increased, each needing to send and receive multiple pieces of information at the same time without interference.

Because smartphones perform numerous tasks simultaneously, each requiring the transfer of information, they have been affixed with multiple antennas, each operating in a different frequency band. Premium smartphones built for 5G can have eight antennas or more. Realme GT 2 Pro, for example, is equipped with twelve antennas! The main communication antenna is called the *cellular antenna* and is used for voice calls and data. Today's cellular antennas operate within the frequency band between 600 MHz and 39 GHz. Cell phones also come with Wi-Fi antennas, which emit and receive two frequency bandwidths, 2400-2484 and 5150-5850 MHz. The Bluetooth capability uses the lower Wi-Fi band (2.4-2.4835 GHz). If the phone has *near-field communication* (NFC) capability, that is the ability to transmit data over short distances (typically 4 cm and less), the NFC antenna operates at 13.56 MHz. This feature has become a staple for smartphones, allowing contactless payments. The lowest-frequency antenna housed in the cellphone is for wireless charging. Qi chargers operate within the VLF frequency band, between 80 and 300 kHz.

Smartphones and other electronic devices often have the ability to receive precise location data from satellites. This has enabled apps to follow your movements and provide navigation. It can be comforting to know you can be found in an emergency, but surveillance can have a dark side in the hands of those who want to stalk or do others harm. In order to track location, smartphones access the *Global Positioning System* (GPS) network, which uses yet another antenna called the *Global Navigation Satellite System* (GNSS) *antenna*. This antenna, however, unlike the other antennas mentioned above, only *receives* signals from satellite transmissions, most commonly at 1.57542 GHz. Although the GPS antenna does not *transmit* signals, depending on an app's settings, location data can be uploaded from the phone to the network, even when apps are not engaged.

Carrier Aggregation

Here is where things get really tricky. Both 4G LTE and 5G phones use a technique called *carrier aggregation*. This is a form of modulation that combines multiple frequencies under one carrier-wave frequency. This technology enabled doubling or even tripling of data transmission rates. The biggest unknown is what frequencies are being aggregated into the carrier frequency. The specific modulation parameters constantly change depending on what is being transmitted. I have read that these modulations include frequencies as low as 10 MHz. Are ELF-EMF or VLF waves *ever* incorporated into the carrier aggregation? Some biophysicists I've talked to say yes. Others say no.

The advanced technologies of carrier aggregation, beamforming, and massive MIMO have all enabled 5G to achieve super-fast data transfer rates. But these technologies have significantly augmented our electromagnetic radiation exposure to a much higher number of increasingly erratic and higher-intensity frequencies of radiation–at close range. This brings with it a corresponding increase in concern about the biological impact of this complex MW and MMW radiation exposure.

CHAPTER 9

RF-EMF AND THE SKIN, EYES, BRAIN, THYROID, TESTES, AND BREASTS

According to the Centers for Disease Control (CDC), people in the US have statistically been living longer lives, with the exception of the past few years when, between 2019 and 2021, life expectancy dropped 2.7 years.[148] Whereas this recent downward trend can readily be attributed to the COVID-19 pandemic, it is also true that the overall health of people in the US has degraded since the start of the new millennium. In 2001, 21.8 percent of adults had at least two diagnosed chronic conditions.[149] Over 25 percent of US adults had three or more diagnosed chronic conditions in 2012.[150] Six years later in 2018, over half (51 percent) of US adults had at least one chronic health condition, and 27.2 percent had multiple chronic conditions.

According to the US Census bureau, the median age of the US population was 35.3 in 2000 and 38.2 in 2018. This steady rise in chronic disease affecting the population, therefore, cannot be explained by an increasing population age.

Medications abound, and it is not only the elderly who are taking a handful of pills each day. Even our children are on antidepressants, antianxiety medications, and drugs for ADHD, among others. From 2013 to 2014, almost 20 percent of our children were taking at least one prescription medication.[151] Why has our health, and particularly our children's health gotten so bad?

Certainly, there are a myriad of environmental toxins negatively affecting the population's health. Electropollution is one of them. Governments have known about the negative health effects of EMF, specifically in the microwave and radio-frequency bandwidths, for decades. Health effects from modulated RF radiation may have been occurring all along. Even radio broadcasting station antennas, which have been in use since KDKA in Pittsburgh went on the air in 1920 and in European countries since the British Broadcasting Company began broadcasting in 1922, have been associated with an increased incidence of melanoma and breast cancer in twenty-three different European countries.[152] In 1977, Paul Brodeur wrote a book called *The Zapping of America: Microwaves, Their Deadly Risk, and the Cover-up*, in which he cautioned the general public about the impending societal health effects if wireless technology were pursued. His book did stoke public interest. In fact, New York City briefly placed a moratorium on the construction of microwave transmission towers. Since that time, research continues to reveal additional negative health effects from this technology.

But, as I mentioned earlier, reports of health effects from MW radiation in the medical literature are often inconsistent. Industry-sponsored studies tend to show fewer adverse bioeffects than studies conducted by independent researchers, suggesting industry bias[153]–and *many* studies are funded by industry. On the other hand, studies employing real-life exposures from commercially available devices are highly consistent in demonstrating adverse health effects.[154] Perhaps some discrepancies are the result of differences in technique. A small change in scientific method can change an outcome. When meta-analyses aggregate research results from industry-sponsored research with that of independent scientists using different protocols, the result can be a misleading claim of "inconclusive" or lack of correlation due to dilution by skewed research. For those who really want to know what the true medical effects are, this inconsistency has certainly been frustrating. Furthermore, some effects may only appear after prolonged exposure or after a significant delay following exposure. This is particularly problematic as performing "provocation studies" likely to elicit symptoms in subjects exposed to MW radiation can be considered unethical, which leaves the field free to expose the general population without knowing what that exposure will do to them.

The proliferation of gadgets and RF radiation-emitting devices has led to indoor and outdoor environments filled with overlapping waves of energy which vary in intensity, frequency, modulation characteristics, and polarity. Radiation exposure is cumulative. Go to an event with hundreds of people all carrying cell phones and you will be exposed to an enormous amount of MW radiation.

Radiation Penetrates

Frequency and intensity, or power density, are both important determinants for assessing health risk. The principles of physics suggest that the higher frequencies of MW and MMW radiation have very limited depth of penetration into the body's tissues. But as radiation intensity increases, the power density at increasing tissue depths will also increase. A 2.45-GHz frequency emitted from a nearby Wi-Fi router or 4G LTE antenna mounted on a building top will produce a continual emission with a highly variable range of power densities, changing over time depending on what the antenna is transmitting. These emissions may produce a maximum power density of 10,000 $\mu W/m^2$ at the skin's surface. This may sound high, but this intensity is well below the maximum permissible exposure limit of 10,000,000 $\mu W/m^2$ in the US, as set by the Federal Communications Commission. In any case, at the *penetration depth*, the power density of this radiation would be approximately 3,700 $\mu W/m^2$. *Higher* power densities of ten to one hundred times that (100,000 $\mu W/m^2$ or 1,000,000 $\mu W/m^2$ at the skin's surface) would equate to intensities of 37,000 $\mu W/m^2$ or 370,000 $\mu W/m^2$ at the penetration depth, respectively. Naturally, higher power densities would also be found *deeper* than the penetration depth with these higher intensity exposures. These are all considered safe exposures by regulators, at least in the US and many other countries.

If it were only this simple to decode our exposure risk, it would be much less daunting. But carrier aggregation, as previously discussed, exposes the body to a modulated carrier wave with many superimposed frequencies, which may be as low as 10 MHz and even possibly lower. Much like a glass prism can separate wavelengths of white light and display a rainbow of color, the body's tissues have the ability to separate out different frequencies of EMR from a modulated wave, or in other words, *demodulate* a wave complex. This means that even if the carrier wave frequency is absorbed by the first few millimeters of skin, the modulated lower frequencies that ride on the carrier wave may penetrate much deeper into the body, all with an intensity equal to that of the carrier wave.[155] Furthermore, in 1914, Brillouin described a phenomenon that may occur when the body is exposed to microwaves that are *pulsed*, which increases their depth of penetration. Brillouin discovered that pulsed microwaves create electromagnetic "precursors," which can propagate through bodily tissues. *Brillouin precursors* have been shown to propagate 15-20 cm through complex layered media by computer simulations.[156]

Because the clinical effects of RF exposure are related to penetration depth, the most sensible way to present this information is to work from the periphery of the body inward. The highest frequencies of WCR may only penetrate a few millimeters into the skin surface, but as penetration depth increases, we risk damage to deeper tissues.

Superficial Structures

EMF and Skin

You may have noticed that if you hold a cell phone up to your ear for an extended period, your ear and cheek get warm, perhaps even hot. Like the cooking that occurs in a microwave oven, the heating effect from a device emitting MW radiation is not limited to the skin's surface. Increases in temperature can cause denaturation of proteins and loss of enzyme function. (Remember, denatured proteins are proteins that have lost their structure, and structure is crucial to correct protein function.) But industry pays close attention to heating effects when creating new products. After all, tissue heating is the *only* biological effect regulators concern themselves with. Because the energy of radiation increases with higher frequency, skin heating is of greater concern for antennas and devices that emit MMW.

When small areas of exposure–between 0.5 and 1 cm in radius–are briefly exposed to MMW, tissue heating is kept under control by the conduction of heat into the deeper subcutaneous tissues. With larger areas of exposure over 1 cm, flowing blood through the area will cool it off through convection. To date, very few studies have looked at the heating effect caused by exposures *longer* than a few minutes to areas of skin *larger* than 1 to 2 cm in diameter,[157] a condition that will surely be common for those accessing the 5G network. To protect yourself, don't put an active cell phone or other device flush against your skin while you are using or carrying it. Laptops shouldn't go on your lap if you have Wi-Fi or Bluetooth antennas turned on. Certainly, if you feel your skin heating up from a laptop or cell phone, turn it off and take a break, or at least distance yourself from the device.

In addition to a heating effect, MMW can be absorbed by the skin and affect most skin structures in the epidermis and dermis,[158] including nerves, blood vessels, melanocytes, and cells that make up the immune system. Skin absorption of MMW is inversely proportional to its reflectivity and dependent on its thickness and, to some degree, its water content. Thinner skin, such as that lining the back of the hand, has a higher reflectance than thicker skin, like that on the palm of the hand, and so absorption will be less.[159] But holding a phone in the palm of your hand allows for optimal absorption of higher amounts of radiation. Perspiration can increase MMW absorption. In addition, coiled sweat ducts in the skin function as tiny helical antennas and also increase absorption.[160] Although it may seem counterintuitive, clothing can increase MMW absorption as well. If one is wearing textiles such as wool or cotton, the absorbed power density from exposure to 26- and 60-GHz MMW frequencies may increase up to 41 percent and 34 percent, respectively.[161] But even with a potential 40 percent increase in absorption while wearing clothing, as compared with bare skin, exposure should not exceed current guidelines.

Melanin is a crucial component of skin that protects the deeper layers of skin from EMR, particularly ultraviolet radiation, by absorbing and diffusing this radiant energy and emitting it back into the environment as infrared radiation, i.e., heat. MMWs have been associated with reduced skin *melanogenesis*, the formation of melanin. Although EMFs can cause the overproduction of reactive oxygen species (ROS) and lead to oxidative stress, a controlled production of specific ROS is a necessary part of melanin synthesis, and this pathway is apparently inhibited by MMW exposure.[162] Although MMW exposure has not been linked to the development of the pigmentation disorder *vitiligo*, it seems plausible that chronic exposure to MMW could eventually lead to geographic regions of pigmentation loss akin to vitiligo, potentially leading to future health ramifications. A study of over ninety thousand patients with vitiligo found a positive correlation between vitiligo and malignant melanoma as well as other skin disorders, including atopic dermatitis and psoriasis.[163]

EMF and the Eye

The visual system is our bodies' most easily observed sense organ. What we all ideally want is clear, sharp visual acuity with appropriate discrimination of colors and patterns. Many conditions can affect the eye and disturb its function. One of those conditions is exposure to MW radiation. Cell phone usage has been shown to cause blurred vision, redness, tearing, and inflammation.[164] But that is not all. MW radiation can affect many of the eye's structures. As waves of light and MW radiation pass into and through the eye, they encounter multiple structures. First is the *cornea*, which provides a clear covering over the front of the eye. The *iris* then controls the amount of light that passes through the *lens*. The lens focuses light onto the back lining of the eye, called the *retina*.

According to Tony Mannarino, MD, an ophthalmologist at the prestigious Wills Eye Hospital in Philadelphia, Pennsylvania,

Research studies have shown that many structures in the eye can be affected by microwave radiation. An elegant study done by the Applied Physics Lab at Johns Hopkins in conjunction with the Wilmer Eye Institute at Hopkins, showed clear non-thermal effects in a primate model. The cornea of the eye was irradiated with levels of non-thermal energy **which resulted in significant and permanent damage easily seen with the wide field specular microscope.** The iris can also be affected. The iris of the eye has a barrier function, similar to the blood brain barrier. Just as EMFs have been shown to cause leakage at the blood brain barrier, **leakage of the iris was clearly demonstrated, indicating there is significant inflammation and loss of vascular integrity in the eye following microwave exposure.** This research study also showed that Timolol, a common beta blocker, and pilocarpine, a medication often used to treat glaucoma and commonly prescribed to temporarily reduce the need for reading glasses, both increased the severity of iris leakage. With Timolol on board, iris leakage occurred at an exposure level between 0.2 W/m² and 1 W/m², which is within current acceptable exposure limits. [Emphasis added.]

It's true. An exposure of 0.2 W/m² is typical for what one would receive close to a smartphone antenna. You may or may not be aware that beta blockers with similar physiological effects to timolol are commonly prescribed to treat hypertension. Unfortunately, no research to date has been performed to determine what effect MW radiation has on the eyes of a person taking beta blockers.

Kues and Monahan, the authors of the study referenced by Dr. Mannarino, summarized their research findings as follows:

The experiments reported herein demonstrate that **exposure to relatively low levels of microwave radiation can result in significant ocular changes** in the nonhuman primate. These changes range from cellular disruption to altered visual function. At 10 mW/cm², **we**

have observed corneal endothelial lesions, increased iris vascular permeability, and degenerative changes in the cells of the iris and the retina.[165] [Emphasis added.]

As if that weren't concerning enough, research has shown that the covering of the lens, a cell layer called the *epithelium*, can also be irreversibly damaged by cell phone radiation at non-thermal levels.[166, 167] A healthy lens is crucial for precise vision.

Retinal effects from MW radiation have been documented, but damage has been observed with higher exposures (*SAR of 4.0 W/Kg*) than those allowed by current safety limits.[168]

Dr. Mannarino's concerns are clear:

The logical conclusion is that over time, exposure to microwave radiation at levels consistent with cell phone use may contribute to cataract formation. We need to remember that, just as the case with brain tumors, cataracts will usually take many years to form. Guidelines that were developed with data from short term studies in the 1980s are not adequate to protect from long-term low-level exposure. Over the span of my forty-year career, whereas removing a cataract in a fifty-year-old patient was an unusual occurrence when I first began my practice, it became a common event over time. The number of patients presenting with ocular irritation and dry eye has greatly increased, the role of microwave radiation needs to be carefully evaluated.

And now there's a high-frequency band of the 5G network consisting of millimeter waves (MMW). Most research studying the effects of MMW on the eye have also focused on potential heating effects. MMW can be absorbed by the cornea and, at very high intensities, have been associated with eye damage in the form of cataracts. This has been documented in epide-miologic studies and case reports in humans. The degree of damage is related to the power density of the radiation and the duration of exposure, exposure which leads to oxidative stress and the deformation of heat-labile enzymes such as glutathione peroxidase.[169]

Pulsed MMW has been shown to create nonuniform cor-neal exposure due to interference effects. Even brief exposures to pulsed MMWs up to ten seconds long can result in localized areas of heating.[170] We have a built-in protective mechanism in that the eye feels pain when its temperature increases 9-11°C above its baseline temperature, and this pain threshold is 10°C below the temperature at which visible corneal damage begins to occur, according to modeling studies.[171]

It is unlikely that a heating effect from MMWs at power densities within existing regulatory limits will cause corneal damage. On the other hand, just as with MW radiation, the potential for long-term, chronic MMW exposure to generate ROS and oxidative stress leading to injury over time is a con-cern and has not been studied.

The Thyroid Gland

The *thyroid* is an organ that sits at the base of the neck, often just below the skin surface. This gland is a storage vehicle for two important hormones called *triiodothyronine* (T3) and *thy-roxine* (T4). Tightly regulated and controlled amounts of these hormones travel from the gland through the bloodstream to have distant effects throughout the body.

The thyroid gland is not supposed to determine when to release these hormones on its own, but rather is to wait until instructed to do so by the central nervous system. The signal is initiated by a part of the brain called the *hypothalamus*, which produces and releases a hormone called *thyroid-releasing hormone* (TRH), which in turn signals the *pituitary* gland to secrete *thyroid-stimulating hormone* (TSH). The thyroid gland recognizes TSH as a chemical signal and, in response, releases its stored T3 and T4 hormones. When all is working as it should, a feedback loop between the hypothalamus, pituitary, and thyroid glands ensures proper levels of circulating thyroid hormones.

Nowadays, it seems that nearly everyone has, or knows someone who has, a problem with their thyroid gland. Either the gland isn't releasing the proper amount of hormone or is filled with nodules. If the thyroid gland is dumping too much hormone into the bloodstream, one may be diagnosed with an overactive thyroid, a condition called *hyperthyroidism*. This diagnosis is associated with symptoms such as heat intolerance, weight loss, and heart arrhythmias. Abnormal thyroid function can also negatively affect the functioning of the immune system.

Inflammation of the thyroid gland, called *thyroiditis*, can cause thyroid levels to initially be too high, but then go down below normal over time as the thyroid gland gets "burnt out," a condition called *hypothyroidism*, which can be associated with weight gain, fatigue, cold intolerance, and depression, among others. Radiation exposure is one of the many causes of thyroiditis. Radiosensitivity of the thyroid gland is well established scientifically. This is why a dentist will often place a thyroid shield over your neck when you have dental X-rays.

Although bioeffects on the thyroid gland were initially described in people previously exposed to *ionizing* radiation, many studies have looked at the impact *nonionizing* radiation can have on thyroid hormone levels. Although a precise mechanism has not yet been clarified, the research is compelling. In one early study, pulsed, modulated RF radiation with a carrier wave frequency of 900 MHz was shown to inhibit thyroid hormone secretion.[172] In 2017, a cross-sectional study performed on eighty-three healthy students showed a significant association between total radiation exposure from excessive mobile phone use and increased serum TSH levels, indicating a hypothyroid state.[173]

In 2019, a literature review was performed by Asi et al. which looked at the effects of wireless communication radiation from 450 MHz to 3.8 GHz on thyroid cells and hormones. Among twenty-two studies, five showed decreased T4 levels, two increased T4 levels, six decreased T3 levels, and one increased T3 levels. Ten of the 22 studies showed alterations in TSH levels from wireless radiation exposure. These authors concluded that MW radiation exposure may negatively influence iodine uptake in the thyroid gland or cause a secondary effect via increasing the temperature of the gland.[174]

A more recent review suggested that cell phone radiation might be associated with thyroid gland insufficiency (hypothyroidism), possibly via disruption in the *hypothalamic-pituitary-thyroid axis*.[175] Therefore, it is possible that EMF may be directly affecting the hypothalamus, the pituitary gland, the thyroid gland, or any combination of the three. I suspect that many of today's cases of abnormal thyroid function are ultimately due to MW radiation exposure.

Time for a Story

A while back, I attended a daylong seminar at the annual meeting for the American Academy of Anti-Aging Medicine (A4M). A group of international scientists presented their research on the health dangers associated with EMF, which at the time I knew very little about. I listened intently to the lectures and left concerned. Although I regrettably did not become an activist at that time, I did make lifestyle changes to reduce my EMF exposure at home and work and suggested my family members do the same.

Shortly after the conference, my sister was diagnosed with hyperthyroidism. When I heard the diagnosis, my initial response was frustration. As a radiologist, I had been reading an increasing number of thyroid ultrasounds on patients with a clinical diagnosis of hyperthyroidism. Even though I was reading for a relatively small community hospital, dozens of thyroid ultrasounds were performed each week on patients. It seemed like everyone had thyroid disease!

My sister was referred to an endocrinologist, who ordered a thyroid ultrasound which revealed several small nodules. Because some thyroid nodules can be "hyperfunctioning" or cancerous, the physician biopsied the most suspicious nodule. Thankfully, her biopsy came back negative.

Because of potential health complications from hyper-thyroidism, including heart arrhythmia, the endocrinologist strongly recommended my sister undergo treatment. Her options, as he saw it, were to either *ablate* (i.e., destroy) the thyroid gland with ionizing radiation or have it cut out by a surgeon. The endocrinologist reassured her that she could take the synthetic thyroid hormone Synthroid (for the rest of her life) and have an otherwise normal life. My sister called me to discuss her options. Neither option appealed to her, or me for that matter. I wanted to know *why* she had this condition before her thyroid gland was killed off.

Then, I asked her, "Where do you keep your Wi-Fi router?"

She replied, "In my bedroom."

I could tell from her quizzical response that my question came across as a non sequitur. Why the heck was I asking her about her Wi-Fi router when she needed to discuss her thyroid treatment options?!

"How long has it been there?" I asked.

She thought about it for a brief moment and replied, "Three or four years."

I explained to her that the Wi-Fi radiation from the router might be affecting her thyroid gland.

"If it were me," I explained, "I would not make a decision about treatment right now. I suggest you move the Wi-Fi router down to the basement, turn off or place your cell phone in airplane mode at night, and have your blood work retested in a few months."

She liked my suggestion and trusted my instinct. Her endocrinologist, on the other hand, wasn't too happy. He wasn't aware of the research showing a connection between Wi-Fi and thyroid disease. But my sister, the attorney, had made up her mind, and there was no changing it.

She moved her Wi-Fi router downstairs, and at her follow-up appointment, repeat blood work was ordered. Her TSH level was now normal. The endocrinologist was confused but said he wanted to follow her with repeat blood work to make sure her thyroid levels remained in the normal range. The results from follow-up blood work were normal. No treatment required.

Her thyroid function remained normal until she moved into an apartment at a fifty-five-plus community four years later. After two years of living in her new home, her TSH level came back elevated, indicating hypothyroidism. The following year, her TSH was abnormally low. I went to her apartment with my EMF detectors and found the levels of RF-EMF in her bedroom to be very high. We unplugged the TV and cable box and turned off cell phones and the Wi-Fi router, which had been strategically placed in a room on the other side of the apartment, but the RF intensity over her bed was still too high. After a little investigation, we learned the neighbor's Wi-Fi router was on the other side of their common wall, and radiation from the router was penetrating through into her bedroom. We set up a Faraday cage in the form of a sleep canopy (which will be discussed in greater detail in chapter 12). And after a year of sleeping in the bed canopy, her T3 and T4 thyroid hormone levels once again normalized. But this time, her TSH levels remained low. Further testing showed she had developed a hyperfunctioning nodule that was ultimately removed.

This story is anecdotal, meaning it's the story of one person/patient. And as such, it doesn't prove an association between RF radiation and abnormal thyroid function, but it certainly correlates with peer-reviewed research published in the scientific literature. Research into the formation of nodules from radiation exposure has not been performed, but structural damage to the thyroid gland follicles has been demonstrated when the gland is exposed to RF-EMF, and the extent of damage correlates with the intensity and duration of radiation exposure.[176]

Reproduction

Occasionally, the news media reports that infertility rates have been increasing. Indeed, infertility rates have been increasing globally since at least 1990.[177] Epidemiologic studies have shown a sharp decline in sperm counts in developed countries.[178, 179] Environmental toxins and toxicants, such as estrogen-mimicking compounds, are known to affect male hormonal levels.[180] The impact of RF and MW radiation on men's health–and in particular testosterone levels and sperm viability–has been an important research topic.

Proper sperm production requires the "factory," the testicles, to be at a temperature below normal body temperature; hence, they hang within the scrotum outside the body proper. The testicles are located within a few millimeters of the skin surface and are therefore easily penetrated by MW radiation. High intensities and prolonged durations of exposure can be common, particularly if a man keeps an active cell phone in his front pocket or frequently uses a laptop computer. MW radiation exposure can increase the temperature within the scrotum and cause oxidative stress within the sperm and testicular cells. ROS can cause DNA damage in sperm, impaired sperm motility, structural anomalies, and decreased sperm count.[181, 182] DNA damage can result in genetic mutations, a scary proposition if someone is looking to have children.

Negative effects on fertility have been described following just four hours of EMF exposure–including reduced testosterone (the "low T" mentioned in numerous pharmaceutical ads), DNA damage, and increased testicular cell suicide (a process known as *apoptosis*)–presumably related to temperature elevation and oxidative stress.[183] Widespread decline of testosterone levels in men has been ongoing for years, which explains the number of male-enhancing supplements in the market.

The takeaway message here for you guys is, in addition to watching what you eat and to exercising regularly, you should turn off your cell phone, or at least put it in airplane mode, before putting it into your pocket. Even if you are not planning to have children, I suspect you probably don't want low testosterone levels, which will not only cause your sex drive to dwindle and your erections to be less than reliable, but also cause changes in your mood, a decrease in your muscle mass and strength, and an increase in body fat. Be aware the same is true for game consoles, laptops, and other wireless devices.

Breast

There may be a correlation between MW radiation exposure and breast cancer in both women and men. Like the thyroid and testes, the breast is a superficial radiosensitive organ. The breast contains glandular tissue that can degenerate into cancer. The incidence of breast cancer continues to rise in developing countries around the world. The list of environmental contaminants potentially linked to breast cancer is extensive and was covered in a 2017 review article in the journal *Environmental Health*, by Gray et al.[184] Among the comprehensive list is nonionizing radiation, which they acknowledge has not been consistently shown in research studies to cause breast cancer in research studies.

However, in 2013, a stunning case report documented multifocal invasive breast cancer in four young women aged twenty-one to thirty-nine years who regularly carried a smartphone in their bra, directly against the skin, for up to ten hours a day over the course of several years. Multifocal cancer means that there were several *different* cancers that developed at the *same time* in the breasts of each woman. In each person, all of the cancers occurred directly beneath the area where she carried her cell phone. None of these women had a family history of breast cancer or positive markers to indicate a genetic predisposition. The specific cancer subtype, *multifocal infiltrating ductal carcinoma* superimposed on *extensive ductal carcinoma in situ* (DCIS), was the same in each case.[185] The odds of this occurring by chance are astronomical. For me, this paper was a smoking gun, not only revealing radiation's link to breast cancer, but implying a link to other diseases as well. No woman should *ever* carry an active cell phone in her bra, nor should men carry one in their shirt or suit pocket–breast cancer occurs in men too. Unfortunately, this behavior is common.[186]

A recent meta-analysis performed in 2020 by Shih et al. looked at 894 health controls and 211 patients with breast cancer and concluded that excessive smartphone use significantly increases the risk of breast cancer.[187] In this analysis, the authors could not calculate how much of this increased risk was directly related to MW radiation causing oxidative stress in breast tissue or how much was due to other hormonal effects. It has been suggested that MW radiation may have a negative impact on melatonin production and secretion, although the research to date is not conclusive.[188] Decreased melatonin levels, from whatever cause, can result in a corresponding decrease in suppression of estrogen secretion,[189] resulting in increased estrogen levels, which in turn increase the risk for developing breast carcinoma. Unfortunately, the Shih paper was retracted due to a reader's concern that two of the references were based on 50-Hz ELF-EMF and not RF radiation, and a third reference focused on male breast cancer. In my opinion and in the opinion of the authors, these were not valid reasons to retract the article, and so I have included their research here.

Despite attempts to squelch public concern about a possible relationship between cell phones and breast cancer, the word is getting out to the public. Rapid dissemination of information is one of the good things about social media, even though misinformation can muddy the water. A friend recently told me that, in the middle of a Pilates class, her instructor emphatically told the class to never put a cell phone in their bra or pocket because it could cause cancer. When I heard this, I felt a small sigh of relief. The public is awakening!

Systemic Effects from Irradiation of Deeper Structures

The Brain

Neurologic effects from nonionizing radiation are not limited to ELF-EMF, but can also occur from MW radiation. When one places a cell phone up to the ear, the intensity of the radiation can be over 2,000,000 $\mu W/m^2$ at the skin surface. While this is a very high-power density, it is still well within regulatory limits.

When the brain isn't functioning properly, it can result in nonspecific, oftentimes bizarre, symptoms. Brain and nervous system symptoms are sometimes hard to "prove," and they may come and go. Symptoms attributed to RF and MW radiation exposure have been reported by numerous researchers dating back to the earliest days of radar. A recent research paper documented that workers exposed to military radar and radio transmission frequencies between 3 and 30 MHz experienced a higher incidence of nervous system symptoms, including headaches, dizziness, and insomnia, symptoms similar to those of *microwave syndrome*,[190] a condition now referred to as EMR syndrome that is fully discussed in chapter 11.

Insomnia

As a younger man, I could sleep anywhere and at any time. I loved sinking into a plush, comfortable hotel bed but could also sleep soundly while tent camping and sleeping in a bag on cool, damp ground.

The first time I spent a sleepless night was at my father's house in 2014. The beautifully decorated guest suite sported an earth-toned color scheme and rich fabrics. Outside were pine trees and the shores of Lake Tahoe. Amid this elegance, a Wi-Fi router had been placed on the nightstand next to the bed. I looked at the apparatus and asked if I could turn it off before heading to bed. My father explained the security system required Wi-Fi, so he preferred to leave it on. Despite my attempt to relax in that comfortable bed, I couldn't fall asleep. My body felt as if it were lying on the beach during a cloudy day. I didn't feel heat, rather a vague sense of being energized. I lay awake the whole night, tossing and turning.

That was the one and only night I slept in that room. Unfortunately, the enjoyment of sleeping in hotel rooms ceased shortly thereafter as well. After the installation of Wi-Fi routers became commonplace, sleeping in hotel rooms became very difficult for me, as well as for many others I have spoken with. The bedroom in my own home even became a place of stressful sleep after a smart meter—emitting nearly continuous pulses of RF radiation—was installed on the house outside my bedroom.

That has been my experience, but what does scientific research have to say about RF radiation and sleep? Although the mechanism explaining how EMR in the visible light band-width can suppress melatonin and affect sleep is clear, the potential impact of lower frequencies of EMR in the RF range has not been delineated. I use the word *potential* because, despite my own personal experience, research linking RF radiation exposure and melatonin levels to date has been conflicting.[191]

Mood

An increasing number of people are suffering from anxiety and depression. RF radiation may be contributing to this condition. Most research studying the effect of RF radiation on mood has been on animal models. One study, for example, demonstrated that rats exposed to thirty minutes of mobile phone radiation every day for four weeks showed a significant increase in anxiety-related behavior. When they sacrificed the animals, the researchers found there was increased oxidative stress in the brain. Vibration, ringtones, or both produced a similar effect.[192] An interesting study from Bolivia documented that rats exhibit increased stress and aggression, along with changes in social behavior, when exposed to EMF.[193]

As you may be aware, rats are often used as a model to study biological effects in humans. After reading these research papers, I began to wonder if ubiquitous RF and MW radiation in our environment has caused increasing aggression in the human population. Certainly, mass shootings, airline violence, road rage, and aggression on streets and in public squares and in schools have dramatically increased in the past twenty years.[194]

Confusion

Some people report difficulties concentrating or thinking when in close proximity to RF-EMF-emitting devices, including cell phone base stations and Wi-Fi routers.[195] But this impact has not been proven. A meta-analysis from 2012 failed to establish a link between cognition deterioration and cell phone use.[196] However, a more recent study performed on rats documented impairment of spatial memory (a cognitive function) after the rats were exposed to twenty-eight days of RF-EMF.[197] Given the importance of a potential effect on cognition and memory, researchers are currently performing a systematic review assessing the impact RF-EMF has on cognitive performance in humans.[198]

As modulation techniques advance, power densities increase, exposure times become incessant, and the increase in multiple overlapping frequencies pollutes our environment, I suspect the effect on cognition will become obvious. Perhaps confusion and other cognition effects may be a secondary effect related to blood clumping "rouleaux formation," previously described, causing diminished oxygen delivery to and carbon dioxide removal from the brain.

Higher frequencies and intensity of RF-EMF have also been implicated as causing typical symptoms of MW sickness, including tinnitus, fatigue, severe insomnia, confusion, headaches, and short-term memory deficits.[199] In a study by Wang et al. in 2017, mobile phone use was associated with a 38 percent increased risk of developing a headache, and the higher the call frequency and the longer the duration, the more the likelihood headaches would occur.[200]

Do Cell Phones Cause Brain Cancer?

Because of the close proximity of the mobile phone to the brain during a call and the ability for RF and MW radiation to cause oxidative stress, scientists and the public continue to be concerned that cell phone use may increase the risk of developing a brain tumor. Like most of my contemporaries, the only time I had ever even heard of a brain tumor was through watching the movie *Death Be Not Proud,* which brought the exceedingly rare condition into mainstream consciousness. Now, however, this type of cancer has become all too common. A member of my extended family, who used her cell phone constantly, developed a brain tumor–the first tumor of its kind in the family's history.

Many animal model studies have been performed to assess the potential for RF radiation to cause cancer. Two important groups have studied this association, the National Toxicology Program (NTP) and the Ramazzini Institute. The NTP examined the effect of RF radiation at 900 MHz and 1900 MHz, frequencies used by cell phones, on rats.[201, 202] The studies found clear evidence of an association between RF radiation and tumors in the heart–called *malignant schwannomas*–in male rats. The study found some evidence for an association with brain tumors and adrenal gland tumors in male rats as well. In the NTP studies, no clear association of tumor development was observed with 900 MHz exposure in female rats. No clear association of tumor development was observed in either sex with 1900 MHz exposure. However, the Ramazzini Institute in Italy performed RF studies of 1800 MHz RF radiation on rats and found that exposure levels much less than those used in the NTP studies increased the risk of heart schwannomas. They also found an increase in brain tumors in female rats with 1.8 GHz RF radiation exposure.[203] These two studies helped raise the alarm that RF radiation may cause cancer in humans.

Skeptics say that if there were an association between EMF and brain cancer, there would have been a dramatic increase in the number of brain tumors since the creation of mobile phones, and this has not occurred. The incidence of *glioblastoma*, a malignant cancer of the brain, *has* increased slightly, but not as much as would be expected if cell phone use caused this tumor.[204] A number of meta-analyses have been performed, but they have been inconclusive. Some researchers have determined there *is* an association between cell phone use and brain tumor development,[205] while others have in a sense kicked the can down the road, stating that epidemiological studies do not suggest an association between mobile phone use and brain tumors, but there is uncertainty about the risk with long latency periods over fifteen years and any increased risk from childhood exposure.[206]

Yes, the likelihood of a latency period between exposure and tumor development *is* a challenging aspect to collecting and analyzing this data. In other words, how many years of repeated exposure to WCR from a cell phone on average does it take before a tumor starts to grow? And how big does the tumor have to get before it starts to cause symptoms that warrant an imaging study such as a CT scan or MRI? Many other factors also impact risk for cancer, including the immune status of the individual, attitude, and the appropriate intake of antioxidants.

There are many other different types of tumors that can occur in the brain and central nervous system that aren't as devastating as glioblastoma. The most common benign brain tumors are *meningioma* and *acoustic neuroma/schwannoma*. Neither of these are actually "brain" tumors, meaning they don't arise from the neurons. Meningioma is a tumor in the tissue covering the brain, and acoustic neuroma is a tumor in the cells covering the nerves that go to the inner ear. Meningiomas and acoustic neuromas grow slowly. Neither tumor will typically cause symptoms unless they grow large and press on the brain. The acoustic neuroma can compress the nerves it covers and can cause hearing loss, ringing in the ears (tinnitus), and dizziness.

Just as with glioblastoma, the literature is all over the board with regard to causation from mobile phone use. In 2006, a study by Schüz et al. looked at the potential effect that RF-EMF emitted from a DECT cordless phone base station could have on tumor development and concluded that these devices do not increase the risk for brain tumors or meningiomas.[207] However, other studies have documented that after a latency period, which can be ten to twenty-plus years, mobile and portable phone use increases the risk of developing an acoustic neuroma.[208, 209] Given that these are slow-growing tumors, a prolonged latency period would make sense. In short, research has been inconclusive.

When a cell phone is placed up against the ear, the radiation travels through the external auditory canal and penetrates the eardrum into the middle ear, gaining access to the brain. The radiation will also penetrate the salivary glands, in particular the parotid gland, which is located a centimeter or so in front of the ear. This is an important gland through which the nerves controlling the facial muscles pass. Salivary glands can develop benign or malignant tumors. A 2019 meta-analysis indicated that salivary gland-tumor risk is not increased in the short term from mobile phone use, but the authors express uncertainty about the possibility of an effect after a latency period of > 15 years.[210] See a trend here?

A recent meta-analysis performed by Choi et al. concluded that cumulative cell phone use of over one thousand hours statistically increases the risk for developing a brain tumor.[211] No one in mainstream media reported on this paper. However, when an analysis commissioned by the World Health Organization was published in 2024,[212] media outlets around the world splashed celebratory headlines like "Researchers Say No Connection Between Cellphone Use and Brain Cancer." Researchers in the field, however, criticized this WHO analysis as being deeply flawed, claiming studies were cherry-picked and other improper scientific methods were used to reach a misleading conclusion.[213]

Research on the impact of RF and MW radiation on the brain has proven to be nebulous and inconsistent. This shouldn't be surprising. If we consider other forms of energy, such as sound, sunlight, and motion in general, people demonstrate varying sensitivities. Some people get sunburned more easily than others. Some get seasick while still moored at the dock. Others can't tolerate loud music. With regard to research on RF and MW radiation, variables such as power density, modulation characteristics, frequency, absorption, duration of exposure, and the underlying health of the individual are pretty much impossible to standardize (or even "control for") when designing a human study. It is likely for this reason and others that research studies have provided inconsistent data.

I know, it's frustrating. I would love to interpret the inconsistent data as "no news is good news" and conclude there is no risk for developing a benign or malignant brain tumor from cell phone use–but I do not. Understanding the basic physics of EMR and fundamentals of the biochemistry of oxidative stress, I don't trust that penetrating the brain with modulated RF and MW radiation is safe. Regardless of the lack of conclusive data, I cannot believe the technology is safe to apply to the head and neck. I therefore never hold a cell phone up to my ear.

Though using speaker mode for conversation used to be quirky, over time, technologies have changed, and habits followed. Fortunately, today fewer people hold cell phones up to their ear to speak and instead use Bluetooth devices, speaker mode, or plug-in earbuds. The younger generation prefer to text and video chat, neither of which involve holding a cell phone up to the head. Yet, although communication has moved in a positive direction, you would be hard pressed to find younger people at the gym, or anywhere in public, without wireless earbuds in their ears. These devices do produce a lower intensity of radiation than a cell phone, but that energy is directed right into the ear canal. And there isn't much separating the middle ear from the temporal lobe of the brain. In fact, the radiation has a straight shot into the cochlea, and from there, it's a short skip and a jump into the nerves that extend into the brainstem.

I get it. There's nothing better than working out with noise-canceling earbuds so you can be in your own world. But please consider that hearing is a delicate sense made possible by tiny hairs called *cilia* in the cochlea of your inner ear. Loud music causes the blood vessels supplying the cilia to constrict, which can cause them to die, leading to a condition called *noise-induced hearing loss* (NIHL). With the radiofrequency radiation the earbuds emit adding possible rouleaux formation affecting perfusion to the picture, you are exacerbating the stress on these cells.

CHAPTER 10

EFFECTS OF RF-EMF ON THE HEART, BLOOD, AND IMMUNE SYSTEM

Investigations into the bioeffects of radiofrequency and microwave radiation have explored many different organ systems, including the autonomic nervous, cardiovascular, immune, and hematologic systems. Research has been ongoing for decades, but data is still patchy and incomplete. Regardless, teams of researchers around the world have made important observations that are helping to provide a better understanding of how electropollution can affect health.

EMF and the Autonomic Nervous System

Nature is perfect in the enormity of its chaos. Everything in the natural world is imperfectly aligned and imperfectly timed. There is variability in everything, no matter how small or imperceptible it may be. Heterogeneity and imperfection keep the mind focused and "in the now." Nothing in our body operates in perfect rhythm like the ticktock of a clock or metronome. There is even variability in the timing between sequential heartbeats. Your pulse can be measured by counting the number of beats in a minute and may remain fairly steady over a period of time. But if you study a tracing of the electric current running through your heart, an *electrocardiogram* (ECG or EKG), you will notice that there is a small variation in timing between each of the individual pulses. *Heart rate variability* (HRV) is the term given to this variation.

Normally, the heart functions relatively autonomously. The initiation of a beat is generated from within the heart. But when the body is under stress, a primitive part of the brain will turn on the autonomic nervous system, which can, among other impacts, override the heart's autonomy, taking over the heart's rate and rhythm. When this happens, the HRV decreases. A decreased HRV is therefore an indication that the body is in an "on guard" state, akin to a fight-or-flight mode. To be clear, the interrelationship of organs and the autonomic nervous system is much more complex than what I am describing. But HRV can be easily measured and used in clinical decision-making. HRV varies throughout the day and with age. But, in general, a high HRV indicates a healthy body and mind, whereas a low HRV can be interpreted as a sign of a poor health status.

Exposure to RF-EMF and/or ELF-EMF has been shown to affect HRV. In a study exploring the cell phone habits of 148 individuals who used mobile phones for over ten years, the researchers concluded that mobile phone use can lower HRV.[214] An additional study of forty-six adolescent students exposed to 1788 MHz at an intensity of 54 ± 1.6 V/m intermittently for eighteen minutes per trial showed significantly increased parasympathetic nerve activity (a component of the autonomic nervous system) and an effect on heart rate variability.[215] However, a study performed by Wallace showed no significant effect on HRV when their group exposed twenty-six healthy young adults to a twenty-six-minute exposure of 900

MHz RF-EMF produced by a mobile phone in a double-blind study.[216] Whether the lack of an effect on HRV was due to the single frequency chosen by this group or the relatively brief single twenty-six-minute exposure time frame is unclear. As with every other systemic effect of RF radiation on the human body, the literature is filled with conflicting data.

The Cardiovascular System

Normally, we are unaware of our heartbeat. But on occasion, people describe their heart as "pounding," "racing," or "skipping a beat." Although these aren't medical terms, they are readily understood, as we have all experienced these symptoms. If experienced occasionally, each of these symptoms can be a normal occurrence. But if they happen with some degree of frequency, they can also raise the suspicion of disease. Cardiologists study the heart, managing vascular disease, blood pressures, abnormal heart rates, and atypical rhythms. They tell us that elevated blood pressure, hypertension, and cardiac arrhythmias such as *atrial fibrillation* have become increasingly common in our population.

The heart contains an electrical system within its muscular framework. These electrochemical conduction pathways are responsible for the generation of coordinated, rhythmic cardiac contraction. Scientists have raised concern that RF-EMF with or without modulation can directly affect cardiac conduction and cause arrhythmias. For decades, their studies have documented cardiac rhythm disturbance in response to RF-EMF exposure. In 1969, a paper described MW radiation effects on the heart, with people who have underlying heart abnormalities being more vulnerable to this effect.[217] Zori Glazer collected publications in 1971 that detailed MW effects on the heart, including disruption in ECG, hypertension (high blood pressure), chest pain, and myocardial infarction (heart attack).[218, 219] More recently, a paper presented evidence that people who live near radar installations have a greater risk of developing cancer and experiencing heart attacks. Those occupationally exposed have a greater risk of coronary artery disease.[220]

In 1997, a review reported that cardiovascular changes, including arrhythmias, have been described in humans from long-term low-level exposure to RF radiation and microwaves.[221] Havas et al. performed a double-blinded provocation study on human participants, exposing them to digitally pulsed (100 Hz) MW radiation at 2.45 GHz. Forty percent of their subjects became hyperreactive and developed either an arrhythmia or tachycardia, indicating a stress response and a generalized upregulation of the sympathetic division of the autonomic nervous system.[222] Wi-Fi routers emit the same frequency, and as one would expect, researchers have also found that exposure to a Wi-Fi router emitting 2.45 GHz can affect heart rhythm, blood pressure, and the efficacy of catecholamines on the cardiovascular system.[223] Sympathetic overactivity and parasympathetic underactivity–as determined by heart rate variability–was also demonstrated in young healthy people following EMF exposure with frequencies of 2.4 GHz and 2.6 GHz.[224] The specific type of arrhythmia was not documented in these studies. And, as is typical for research in this field, not every study confirms these results.[225]

Some researchers have taken a closer look at specific impacts WCR can have on the echocardiogram (ECG) tracing. The ECG is a tracing that records in real time the voltage generated by the heart's conduction system, which transmits throughout the body. The electrical current generated by the heart can be picked up by leads stuck to the skin. The typical ECG tracing is divided into segments, intervals, and waves (Figure 10.1).

The waves in the graph represent the direction and intensity of electrical activity and are designated by the letters from P through T. Because the Q, R, and S waves occur one right after the other, they are often grouped together and referred to as the *QRS complex*. Any change in the size of a wave or a change in the ECG tracing indicates a change in the electrical activity of the heart. Changes in the timing of intervals can have implications for heart health. One study demonstrated that short-term exposure to *pulsed* ELF-EMF–which some think may occur within modulated MW and MMW radiation–can affect the properties of ECG signals, including mildly increasing ECG voltage levels, and eliciting a small change in the *RR interval*, that is, the time between two consecutive cycles. A change in the RR interval has a reciprocal relationship to the heart rate. In other words, when the RR interval increases, the heart rate drops, and when the RR interval shortens, the heart rate increases.[226]

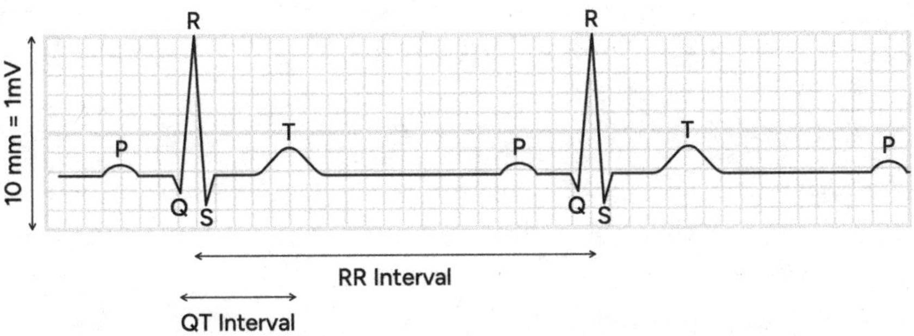

Figure 10.1. ECG tracing

An additional study demonstrated prolongation of *the QT interval* in males with myocardial ischemia when they were exposed to an active cell phone for forty seconds. Interestingly, females

did not demonstrate this effect.[227] An increase in the RR interval and prolongation of the QT interval are both concerning
because they indicate that exposure to MW radiation has an
impact on the heart's conduction system. However, arrhythmias have not *conclusively* been proven to be induced by
WCR.[228] Although arrhythmias have sometimes not occurred
in patient populations under study, it is possible RF radiation
may act as an environmental cofactor exacerbating other factors that might lead to arrhythmia in some people; in other
words, MW radiation may be the "straw that breaks the cam-
el's back."

We are in the midst of an atrial fibrillation (Afib) epidemic,[229]
and the cause is unknown. Could it be related to smart meters?
Maybe . . . I personally believe it is likely to be a contributing
factor. I've heard many people say they developed Afib after
a smart meter was placed on their home or a cell tower was
placed nearby. Could the increase in Afib be due to our overall increasing cumulative exposure to WCR? Yes, that's possible too. It's also possible it could be from something else
altogether. There are many toxicants that could be contributing to the incidence of arrhythmia. For example, glyphosate,
the active ingredient in Roundup®, is a metal chelator. Metal
chelators bind to metals and make them unavailable for use.
Because glyphosate is in the food supply, it may be contributing to decreased supplies of magnesium, leading to coronary
artery spasm and arrhythmias. One thing I am fairly certain
is that the increase in Afib is not merely because the population is aging. In my day-to-day practice, I am seeing younger
and younger patients with cardiac pacemakers. It is no longer unusual to see a pacemaker in a thirty- or forty-year-old,
something that was extremely rare twenty-five years ago
(Figure 10.2).

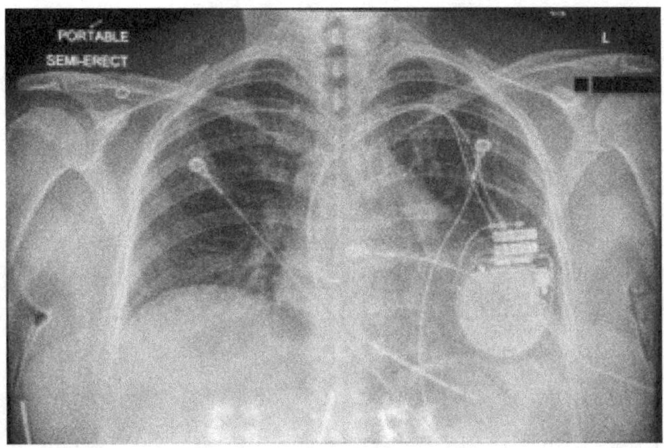

Figure 10.2. Fourty-year-old female with a cardiac pacemaker

Only a quarter to half the people who have received treatment for Afib have pacemakers. A procedure called a cardiac ablation has become a common alternative to a pacemaker and may be performed in some people, relieving them of the need to walk around with a cardiac pacer. In an ablation, a piece of the heart–the AV (atrioventricular) node–is burned up (or frozen, *cryoablation*) so it doesn't function anymore. This stops the arrhythmia from occurring. An ablation procedure is performed by placing a catheter into the heart from which focused RF radiation is aimed at the node.

Interestingly, for those who need a *cardiac implantable electronic device* (CIED) like a pacemaker, newer generations of common technology can potentially cause problems. Some cell phones and accessories–including smartwatches and wireless headphones–have the ability to interfere with CIEDs. Because of the stronger magnetic fields built into this equipment, if the phone or accessory is held within six inches of the implanted device, it can cause electromagnetic interference, induce a magnetic response, and cause the cardiac pacer or defibrillator to reset.[230] Other smart devices, including smart scales and smart rings, may also interfere with implantable electronics.[231] Just in case you were wondering, airport security scanners, which emit unmodulated frequencies between 70 and 80 GHz, do not have any effect on implantable cardiac transvenous pacemakers.[232]

Arteries

Arteries deliver blood from the heart, normally becoming narrower and narrower the further they go. These blood vessels are lined with a thin layer of cells called the *endothelium*, which provides a smooth lining for the passage of blood. When the endothelium becomes damaged, trouble can ensue. Oxidative stress leads to endothelial damage.[233] Excessive free radicals within a cell can result in a cascade of reactions that destroy the cell's integrity. The body's immune system then attempts to heal the damage, which can result in chronic inflammation. A damaged endothelium predisposes the artery to develop *atherosclerotic plaques*, which can, over time, narrow the vessel and impede blood flow. This can result in organ damage, such as a heart attack, a stroke, or even a cold, pulseless leg that could potentially require amputation.

Arteries and arterioles (small-caliber arteries) are at different depths throughout the body and so an individual vessel's vulnerability to MW and RF radiation is relative to the frequency and intensity of the radiation that would reach it, as is also true for the blood cells passing through the vessel. The endothelium is no different from other cells in its response to oxidative stress from EMF, and it seems possible that the increase in cardiovascular disease may be in part the result of this damage. It is well known that ionizing radiation causes endothelial damage. But a study by Xu et al. showed that endothelial damage can also occur from MW exposure, and the extent of damage is dose dependent.[234] But exposure intensities used in this study started at 100 W/m, higher than the maximum exposure limit in the US, which is set at 10 W/m². Regardless, this doesn't mean that lower-intensity exposures *don't* cause damage. It suggests more research is needed.

I have several close friends who had no risk factors for coronary artery disease, had low cholesterol levels and normal blood pressure, and ate healthy diets, yet developed critical narrowing of their coronary arteries in 2022. Although their cardiologists had no explanation for how this could occur, researching this science makes me concerned their disease may have been related to MW radiation exposure, if not as a direct contributor to endothelial damage, then as a cofactor. These stories are anecdotal, but when combined with previous stories I've shared, they raise questions.

Blood and the Immune System

The term "blood" does not refer to a specific type of cell. Rather, blood consists of a hodgepodge of different cell types, proteins, ions, and fluid. Each constituent has its own important purpose. White blood cells are immune cells that fight infection. Red blood cells (RBCs) facilitate the delivery of oxygen to the body's tissues and remove carbon dioxide buildup. Platelets have a role in blood clotting. The cells that comprise blood are unique in that they circulate throughout the body's arteries and veins, reaching peripheral capillaries within the subcutaneous tissues as well as engorging the deep organ systems. RF or MW radiation could, therefore, cause a physiologic effect on the blood in superficial or deep tissues, depending on the frequency and intensity of the radiation and the corresponding penetration depth, but effects on the blood flowing through superficial capillaries could also have an effect on deep tissues that cannot be directly reached by the radiation.

Oxidative stress is likely the underlying mechanism behind the RF/MW radiation impact on blood, just as it is with other cells in the body. In support of this assumption, a study in which humans were exposed to cell phone radiation reported increased blood levels of lipid peroxide, an end result of oxidative stress. The enzymatic activities of two important enzymes–*superoxide dismutase* and *glutathione peroxidase*–in RBC samples were also decreased, further indicating oxidative stress.[235] Another study analyzed the blood plasma of individuals residing near mobile phone base stations and showed they had significantly reduced glutathione, catalase, and superoxide dismutase levels over unexposed controls. These are enzymes responsible for reducing ROS, and their depletion corroborates that MW radiation causes oxidative stress in the blood.[236]

Any damage to the white blood cell line by WCR can impact the normal functioning of the immune system. Indeed, studies on laboratory animals have shown that even low levels of RF radiation can impair immune function. The Russian scientist Grigoriev concluded in 2012 that exposure to MW radiation, even at low levels (0.1-0.5 mW/cm^2), can impair immune function, cause physical alterations in immune cells, and degrade immunologic responses.[237] This power density converts to a value of 1,000,000-5,000,000 µW/m^2, which, as we will soon learn, is well within the range of exposure intensity you receive when holding a cell phone or placing it up to your ear. RF-EMF can cause T-lymphocyte suppression.[238] Rats exposed to 2.45 and 9.7 GHz for two hours a day, seven days a week for twenty-one months showed a significant decrease in the levels of lymphocytes and an increase in mortality at twenty-five months in the irradiated group.[239] Interestingly, rats irradiated with 2.45 GHz at 0.5 mW/cm^2 for seven hours daily for thirty days experienced autoimmune reactions.[240] Although this study was performed using an animal model, I find the results of this research particularly disturbing considering how common autoimmune diseases have become.

In a literature review performed in 2009, Johansson concluded that EMFs can disturb the immune system and cause allergic and inflammatory responses at exposure levels significantly lower than current national and international safety limits.[241] Similarly, a review conducted by Szmigielski in 2013 concurred that even weak RF/MW fields emitted by cell phones can affect immune functions.[242] The reason I use the term *affect*–as opposed to a more specific term like *degrade* or *enhance*–is because the observed impacts of RF on the immune system have been shown to be variable. In some cases, the immune system is enhanced or stimulated into action. In other studies, the effect is inhibitory. In general, experiments seem to indicate that short-term exposure to weak MW radiation may temporarily *stimulate* an innate or adaptive immune response, but prolonged irradiation *inhibits* those same functions. I would consider this to be a "learned helplessness" response of our immune system, analogous to the phenomenon described by the psychologist Martin Seligman, PhD, in 1967. Although the immune system is temporarily spurred into action to defend against this energetic intruder, as time goes by and the assault persists, the immune system becomes tired of the battle, stops fighting, and becomes weakened and deficient in mounting a response.

MMW can affect peripheral circulating immune cells. A literature review and meta-analysis using artificial intelligence techniques showed that increased duration of exposure to EMFs generated by 5G devices decreases immunity and may have increased respiratory complications in patients with COVID-19.[243] In fact, many features of the COVID-19 pandemic, detailed in a paper I cowrote with a biophysicist, Dr. Beverly Rubik,[244] could have been exacerbated by the worldwide presence of RF-EMF. At the time, we analyzed images available on Wigle.net, a website that demarcates the location and concentration of wireless networks around the world, from 2019. We correlated the map with an image from the

Johns Hopkins Coronavirus Research Center from April 2020 that showed the pattern of early dissemination and number of COVID-19 cases in each geographic region. The similarity of the two maps was eerie and disturbingly apparent. Just to see if the relationship might have also been related to population density we then correlated the two maps with an atlas showing population density as of December 2019 obtained from worldpopulationhistory.org. Surprisingly, the wireless network distribution and early COVID-19 dissemination did *not* correlate with population density. I could not obtain permission to display these images in this book, but if you have the wherewithal to check out these sites, you might be able to see these maps for yourself.

The effects of RF/MW radiation on the immune system are not limited to immune dysfunction, but appear to increase the risk for developing cancer of immune cell lines. An in vitro study performed by Panagopoulos in 2019 exposed lymphocytes to thirty minutes of radiation at currently acceptable limits by 3G UMTS (1900–2200 MHz) and ELF-EMF pulses at 100 and 1500 Hz, during the G2 phase of cell division. Panagopoulos then observed that the cells exhibited a metaphase (an important stage of cell reproduction) and gaps and breaks in the DNA at highly significant percentages (275 percent as compared with controls).[245] What this means, in a nutshell, is that if an immune cell is exposed to RF/MW radiation at a specific time during its reproductive cycle, the radiation can damage and break apart the DNA. DNA damage can potentially lead to genetic mutations and cancer.

Supporting this observation, a recent study on military personnel exposed to high-intensity RF radiation for one to three years showed an increased risk for hematolymphoid cancers from 22.7 percent to 41.3 percent after a median latency period of 4.6 years.[246] Because of this significant finding, these authors suggest that the International Agency for Research on Cancer (IARC) change the classification of RF-EMF from a group 2–a *possible* human carcinogen–to a group 1 designation–a *known* human carcinogen.

An observable and interesting transient phenomenon has been shown to occur when RBCs are exposed to RF/MW radiation. As described earlier in chapter 3, by using dark-field microscopy and a technique called live blood analysis, after radiation exposure, red blood cells in some people have been shown to stick together, one on top of the other, resembling a stack of coins, referred to as a rouleaux formation.[247] The phenomenon is reproducible, meaning it can be demonstrated repeatedly in the same patient. It is also a transient phenomenon, in that the effect dissipates and no longer occurs after a period of time, which may be as brief as twenty minutes after radiation exposure has ended.

Aside from looking strange, rouleaux formation can have health effects. The abnormal RBC clumping and aggregation can lead to the formation of blood clots.[248] In addition, it has been suggested that RBCs in rouleaux are less efficient at carrying oxygen.[249] Reduced oxygen delivery can cause symptoms of fatigue and weakness, similar to anemia.

Coherence

The human body is not a collection of separate unique organ systems that work in isolation. Every living being is able to exist because the cells that make up its structure are coherent. All bodily systems support one another. Even though we have divided up medical delivery into what might seem like a systems-based approach–i.e., gastroenterology, neurology, cardiology, etc.–that doesn't mean these systems aren't interrelated. For example, a properly functioning immune system is critical for health of the entire body. An out-of-whack immune system can lead to a greater susceptibility to infectious diseases, like COVID-19 and others. But it can also lead to an increased risk for developing cancer, not only of its own cells, but also those of other tissue types. Autoimmune diseases are also the result of immune dysfunction. An improperly functioning thyroid gland can affect the functioning of the immune system's cells. While at the same time, it can also affect metabolism and cause weight gain or loss.

A Nonselective Agent

By focusing the previous two chapters on a few important specific organ systems, my intent has been to provide a summary explanation of how RF/MW can impact some of our internal workings. But this has not been a complete picture. The body is complex. One may read through the previous two chapters and be disappointed that much of the research on the effects of RF and MW radiation, to date, is not conclusive. And whereas this is true, the effects of RF/MW radiation on the immune system alone should be enough to scare anyone. Did the pandemic *just happen*? Or did the collective immune system dysfunction of a suitable host population allow this novel disease to flourish?

The unfortunate truth is that the effects of RF/MW radiation are not selective. EMF seems able to cause oxidative stress in any cell within the body. Because we are dealing with a form of energy and not matter, EMFs can be absorbed at specific resonant frequencies unique to certain molecules, but they do not respect cell boundaries or protective barriers like the blood-brain barrier and other tissue planes.

In order for a cell to be impacted, it needs to be exposed to the waves of radiation. Frequency and intensity are both important determinants for penetration depth. But with new techniques of carrier aggregation, even high-frequency MMW that may only penetrate a few millimeters into the skin may have lower-frequency modulations within their signal that can be demodulated by the body's tissues and reach deeper structures at high-power densities. A study investigating the effect of modulated 1.8-GHz RF radiation on trophoblast cells–the type of cell that makes up the placenta–concluded that modulated MW radiation can cause DNA damage, whereas an unmodulated continuous-wave of the same frequency does not.[250] Modulation and pulsation of the carrier wave are important characteristics that differentiate wireless communication from other technologies that emit non-modulated RF and MW radiation.

Musculoskeletal Effects?

Some biological effects from RF/MW radiation cause symptoms, while others do not. Unless there is some type of symptom, it is hard to know that anything is going wrong inside the body. DNA damage, for example, won't produce symptoms–at least not right away. It may result in a cancer showing up years down the road, but it will usually be difficult to correlate the formation of cancer with a specific exposure.

Aches and pains, however, are usually front and center in the minds of most people. Many with arthritis can correlate certain types of climates with an improvement or a worsening of symptoms. I have heard several stories from people who were suffering from hip pain that seemingly came from nowhere. These people had no history of injury, abnormal movement, or increased activity to explain their symptoms. One person described the pain to me as "a pain that cannot be ignored and causes me difficulty in rolling over onto my side while I'm trying to sleep." Interestingly, the pain in each case correlated with the side on which a device, such as a smartwatch, was worn. I recommended that these patients stop wearing their device to see if it might be related. In each case, the pain resolved quickly. Although I have not come across any research articles on the association of pain, tendinitis, or bursitis with RF radiation-emitting devices, it seems likely this is another potential injury from this technology. I don't recommend wearing any RF/MW-emitting device directly on the body, including a smartwatch, unless it is kept in airplane mode with all antennas disabled, which kind of defeats the purpose.

In some people, exposure to RF/MW can cause a general "upregulation." Hypersensitive individuals can, in a sense, become allergic to EMF. A recent study performed on rats simulated 5G multifrequency exposure and demonstrated a functional reactivity of an important neurohormonal mechanism called the *hypothalamic-pituitary-adrenal (HPA) axis*. This impact was determined to be a stress response.[251] The HPA axis is a complex system responsible for maintaining balance in the body, a term called *homeostasis*. A disruption in the HPA axis can lead to significant health repercussions, as it can negatively affect neuroendocrine, behavioral, autonomic, and metabolic functions.[252] In the next chapter, we will take a deeper exploration into a population who may be experiencing this phenomenon.

CHAPTER 11

ELECTROHYPERSENSITIVITY SYNDROME

As we keep increasing ambient levels of EMF, it is likely that more and more people are going to become electrosensitive. Those who are my age or older remember a time when few people had allergies. Almost no one was allergic to peanuts, and there was no celiac disease. Environmental contaminants have certainly affected the health status of the population. In previous decades, the population was less anxious and/or depressed, and few people complained of insomnia. I suspect we are on the precipice of a world in which many people are going to become highly electrosensitive. When they do, those not yet afflicted with the condition will not "get it."

A Suffering Without Compassion

When a patient, friend, or family member complains of an "unseen" assailant, such as an environmental toxin, disturbing their health, it is unfortunate that those with whom they share their condition often dismiss their symptoms as being psychosomatic or "in their head," particularly if they have not had a similar experience. This is nothing new. Mysterious illnesses brought on by environmental exposures have been misunderstood for centuries. For instance, ergot poisoning happened numerous times in numerous countries causing physical symptoms and bizarre behavior in many; some scientists and historians even think ergot poisoning triggered the Salem witch trials of the 1690s.

By bringing awareness to the patient as to how their personality or seemingly unrelated earlier life experiences might explain current sensitivities, psychiatrists and psychologists trust symptoms can be explained away and dissipated. This may be a valid approach for some conditions, but the bodies of millions of people in the US alone are now suffering symptoms from trying to defend against unseen toxins, including ELF, VLF, RF, and MW radiation.

I understand this well because my mother suffered a form of environmental sensitivity. The focus of her life turned into a continual search to get well. Being divorced by her husband in the 1960s and left with three young children was certainly a reason to need the support and guidance of a psychiatrist, but her symptoms had begun before our father left and persisted for decades. It was difficult growing up in a home in which our mother was often in her bedroom lying down with the door closed and the lights out because she felt tired and "woozy." She was often irritable and hyperreactive. I spent a good bit of my adolescence outside of the house and tried to be quiet when I entered, just in case she was lying down.

My mother had been an active young adult and tremendously intelligent. In fact, she graduated from Pembroke College, the sister college to Brown University, one of the eight Ivy League colleges. She played classical piano and memorized pages and pages of sheet music, entertaining herself, and us, by practicing works by Chopin, Beethoven, and others. Without a clue as to why she felt ill most of her adult life, she attempted many treatments in the form of allopathic medications and alternative treatments, including diet modification, in an attempt to get well.

She suffered not only from her symptoms, but equally from the dismissal of her symptoms and condition by family and friends. Those in her "support network" diagnosed her as being (1) a difficult person, (2) depressed because her husband left her, (3) just looking for attention, or (4) some other self-serving reason. Whenever the extended family got together, invariably a part of the schedule would be discussion of my mother's "supposed" poor health and ways in which her condition could be improved.

"She needs to get a job."

"She needs to move back home so her family can look after her."

"She needs a man!"

When a neurologist diagnosed her with *neurasthenia*, she felt she had finally found a doctor who understood her suffering. From then on, she had ammunition. When the family told her all she needed to do was go out and get a job, she would use her diagnosis of neurasthenia to explain why she couldn't work. She never knew if she was going to feel well enough to get up and go to work, let alone be able to think clearly for a prolonged period of time at a job. Relatives might have thought she was now hiding behind the diagnosis given to her by a "quack" doctor, but I lived with her and knew her complaints were legitimate.

It wasn't until I left for college and she left the house we had been living in for almost twenty years that she seemed to renew and come alive again. She moved to Florida, where she began to live an outdoor lifestyle. Her energy levels improved, and she slowly became more resilient. She played bridge with her friends most days and even spent time with a boyfriend. Family and friends all noticed the dramatic improvement in her well-being. At the time, no one could figure out what had caused her improvement. "Maybe it's the sea air," she had suggested herself, but most figured she was feeling better because she was living a more active social life.

Why have I recounted memories of my mother's health challenges? In part, I want to convey how important it is to understand that our mental health and physical condition affect others in our lives. A person's infirm health status can affect others within and beyond the immediate family. My mother's best friend from high school suffered continual rejection because my mother often didn't feel well enough to talk to her on the phone. Her brother felt rejection because she didn't feel well enough to have him come visit. Many other friends and family members felt bad for her because they remembered what she was like when she was well and couldn't understand why she was unable to pick herself back up and get on with her life.

Even after extensive research on environmental toxins, I do not know which contaminant my mother suffered from, but I am quite certain her symptoms were the result of hypersensitivity to an environmental toxin. I watched it. I lived with her and the condition she playfully called "it" for a long time. Over the years—and even to this day, years after she passed—I have often mused about what environmental triggers could have caused her weakened state. Perhaps it was mold toxicity. She had a clothes storage closet below her bedroom that had flooded at one point, and all of her clothes became moldy and needed to be thrown out. Certainly, chronic mold toxicity could have accounted for some of her symptoms, although mold toxicity was not a recognized diagnosis at that time. Today, we recognize a condition called *multichemical sensitivity*, which would have certainly explained some of her symptoms. What

I find curious, though, is that she was diagnosed with neurasthenia, which was an early term given to those who suffered symptoms from exposure to electromagnetic fields. In medical school in the late 1980s, I was taught that neurasthenia was a bogus diagnosis, which I am now ashamed to have once believed.

There was no Wi-Fi back then and most people had never seen a cell phone, but we did have electricity running through the house, producing ELF-EMF. In addition, we had dimmer switches to control lights, which we now understand produce dirty electricity. There were also a large number of powerful radio station antennas in the New York metropolitan area. Now, I wonder if her symptoms may have at least in part been related to EMF.

What Do We Know and Understand About the Environment's Effect on Health?

As a society, we have become much more cognizant of how our bodies interact with the environment. It has taken, in my opinion, a ridiculous amount of research and time for scientists to prove simple concepts, like the fact that we interact with the air we breathe, the food and water we consume, and the materials we apply to our skin and absorb. I use the word interact, because it is accurate. We don't just inhale air. The air we breathe in interacts with the blood cells flowing through the tiny capillaries in our lungs. Hemoglobin molecules in the red blood cells exchange carbon dioxide for oxygen molecules with each breath. Other gases we inhale can also infuse into our bloodstream and disperse through and affect our body.

The water we drink doesn't just flow into our stomach and out of our bodies when we urinate. Water affects the movement of electrical current within our cells and organ systems and influences the interaction of molecules. If water has contaminants in it, they too can become absorbed and affect the workings of our body. We are inextricably engaged with the environment. But most of us are unconscious of this relationship.

An understanding that air pollution, including cigarette smoking, can cause chronic heart and lung disease led to the slow societal movement away from this popular habit. People are becoming more aware of fragrances and odors, knowing that they can be damaging to the body. The understanding that we absorb chemicals through our skin has had a major impact on the chemical, pharmaceutical, and cosmetics industries, among others. We now know that applying certain chemicals to the skin results in absorption, possibly causing hormonal disturbances and other health problems. Population groups "in the know" are slowly shifting to safer products.

Getting people to understand that the invisible energy currents emitted by wireless devices are also affecting health is the next major hurdle. But this is proving to be a most challenging task because energy is so abstract and intangible. Understanding the effects of "toxic" energy requires a new construct or paradigm shift in understanding. I believe we are on the fringes of this breakthrough, in part because more and more people are getting sick, and not sick from ordinary processes we understand, but for unexplained reasons. We have witnessed athletes suffer heart attacks for no apparent reason. Reporters have collapsed mid-broadcast. I know of two healthy young women who suffered heart attacks and, when they went for cardiac catheterization, had no plaque in their coronary arteries. Thirty-year-olds now have an increased risk for Afib, an arrythmia that used to be rare.

Awareness

Long-term COVID, the COVID-19 pandemic in general, and vaccine adverse effects have clouded the picture, making it more difficult for people to see a correlation between the new influx of symptoms and the recent proliferation of wireless communication radiation. It is human nature to lose awareness of something one is continually exposed to in the environment. Furthermore, when toxins are invisible, ascribing symptoms to another cause, one that is seen or felt, is common. I believe this is what has been occurring with EMFs. Although many people in our society are actually experiencing symptoms related to exposure, they don't recognize it.

Like most, I had been unconsciously "amped up" from incessant EMF exposure. But it wasn't until a recent trip to Africa that I had this dramatic realization and remembered how I used to feel. My family and I traveled to a remote park in western Tanzania to see wildlife in a relatively non-touristy park. The park was over 300 square kilometers (116 square miles) and we were three of only five tourists in the park that week! When I first entered my tent, furnished with a double-wide cot, I lay down and took a deep breath and sighed. My body immediately started to decompress. I hadn't felt this type of relaxation in many years. Yet, it was a familiar feeling, one I had forgotten. There was no Wi-Fi or cell service in our camp. I slowly drifted off to nap, feeling safe and secure, listening to hippos grunting in the background. At one point during my stay, I took out my RF-EMF detector and turned it on. For the first time ever, I saw a value of 0.0. It was sad to think that I had to travel halfway across the globe to be able to relax outdoors and be in an electrically clean environment. After four days in western Tanzania, I felt healthy, relaxed, and coherent! My blood pressure was no doubt normal too. I dreaded the idea of going back into the world of EMFs. After our three-week vacation, when I did head back home, I felt well for a few days, but soon thereafter, the subtle annoying sense of agitation and anxiety insidiously returned.

Diagnosing Electrohypersensitivity Syndrome (EHS)

In 1991, Bill Rea, a physician and fellow of the American College of Surgeons (FACS), and his coauthors named the collection of symptoms experienced by some people when exposed to EMFs *electromagnetic field sensitivity*. The list of symptoms Rea and others encountered in patients who were sensitive to EMF were remarkably similar to those ascribed to radio wave sickness in early reports written by Russian scientists studying MW energy. Neurologic symptoms such as fatigue, headache, and difficulties maintaining concentration occurred in this population, in addition to dysesthesias, a condition in which senses, such as the sense of touch, are distorted. Many of these patients also described bone and joint problems, cardiovascular abnormalities, and skin abnormalities such as redness, tingling, and burning sensations.

The term *electromagnetic field sensitivity* was changed to *electrohypersensitivity syndrome* (EHS) in 1997. EHS was defined as "a phenomenon where individuals experience adverse health effects while using or being in the vicinity of devices emanating electric, magnetic, or electromagnetic fields (EMFs)."[253] At the time, EHS was used both as the name for a medical condition as well as a descriptor for people with the ability to perceive or react to EMF at significantly lower levels than most people.

People with EHS have described moderate and even severe health consequences when they are exposed to ELF-EMF, RF-EMF, or both, which have often been regarded as having a "psychological" origin. Cognitive and behavioral therapy are frequently recommended as treatment options with varied success.[254] The dilemma for the psychologist is that when these people are removed from EMF, their symptoms go away. For decades, scientists have been trying to understand this phenomenon. Are these people experiencing a psychological disorder, a physiological disease, or are they actually healthy and highly functioning people who are simply being alerted to the presence of an electromagnetic environmental toxin through symptoms?

Some scientists believe that people who experience negative health effects from EMF exposure are experiencing a *nocebo effect*. This term represents the opposite of a *placebo effect*. In other words, instead of a person believing that a treatment is going to heal or help them, they fear or expect an agent is going to harm them. And therefore, it does. A nocebo effect would imply that a condition occurs because of the negative expectation that exposure will cause harm. Many studies have raised the question as to whether or not EHS is the result of a nocebo effect, and some have determined that this could provide a reasonable explanation for symptoms in patients with EHS.[255] Some studies, however, only exposed their test subjects to a limited time frame–50 minutes of EMF exposure–to determine whether or not symptoms developed. This may not have been a sufficient time frame to uncover a systemic effect. Another interesting provocation study performed on sixty-five healthy participants concluded that sensational media reports can sensitize people to develop a nocebo effect to EMF.[256] Yes, these sometimes debilitating symptoms have been the media's fault all along!

While some scientists have painted this condition as a psychological phenomenon, others have looked for physiological responses to EMFs that can be objectively quantified. One of the first, if not the first, demonstration of a measurable and reproducible physiological effect was performed by a scientist in Sweden named Dr. Olle Johansson. He observed skin reactions in the Scandinavian population characterized as inflammation brought on by exposure to radiation emitted from video display terminals. He termed this "screen dermatitis." His provocation study showed those affected with this condition had a high-to-very-high number of somatostatin-immunoreactive dendritic cells as well as histamine-positive mast cells in skin biopsies before exposure. But after exposure, the somatostatin-positive cells disappeared while the mast cells remained, unchanged.[257] This research led the way to other studies designed to discover other physiological effects from EMF exposure. We have seen many of these effects discussed in previous chapters.

A French team, led by Dr. Dominique Belpomme, presented a series of abnormal lab results they observed in anywhere from 14 percent to 40 percent of the EHS population.[258] For interested clinicians, they are listed in Table 11.1.

Marker	Normal Value	EHS Mean ± SE	Above Normal (% of subjects)
Low-grade inflammatory markers			
hs-CRP	‹ 3 mg/L	10.3 ± 1.9	15
Histamine	‹ 10 nmol/L	13.6 ± 0.2	37
Ig-E	‹ 100 UI/mL	329.5 ± 43.9	22
Hsp27	‹ 5 ng/mL	7.3 ± 0.2	25.8
Peripheral blood levels			
S100B	‹ 0.105 µg/L	0.20 ± 0.03	14.7
Nitrotyrosine	› 0.9 µg/mL	1.36 ± 0.12	29.7
O-myelin	Qualitative test	Positive	22.8

Table 11.1. Abnormal lab test results in patents with EHS

Adapted with permission from: D. Belpomme and P. Irigaray, "Electrohypersensitivity as a Newly Identified and Characterized Neurologic Pathological Disorder: How to Diagnose, Treat, and Prevent It," *International Journal of Molecular Sciences*, 2020; 21(6): 1915.

Although each of these abnormal blood values were present in less than half of the study population reporting EHS, the authors also presented data that nearly 80 percent of patients with EHS present with increased oxidative/nitrosative stress-related biomarkers in blood plasma.[259] Elevation of these biomarkers indicates that cells have been subjected to an increased level of oxidative stress.

Even though these blood tests may indicate oxidative stress in patients with EHS exposed to EMFs, correlating exposure to symptoms is another leap. Proving that symptoms in patients with EHS are in fact *caused* by EMF and not another environmental toxin or a nocebo effect, or a combination of both, has been challenging. It has been a very slow process. Scientists have been trying to "prove" a connection between EMF exposure and symptoms in those with EHS for decades. Despite using multiple approaches, results have been inconsistent and insufficient to convince the skeptics or, more importantly, force government regulators and industry to change direction.

Because of the inability to prove causation, the term *electrohypersensitivity syndrome* was called into question at a WHO workshop in 2004.[260] The champion behind this change, Herman Staudenmayer, PhD, an attendee and psychologist from the US, expressed concern that the term suggested cause and effect. In its place, he recommended the term *idiopathic environmental intolerance* (IEI) to electromagnetic fields. This term had first been introduced in 1996 at a workshop in Berlin held by the WHO's International Program on Chemical Safety as a more suitable descriptor for a condition with a self-reported symptomatic reaction to an environmental trigger without proven causation or defined pathophysiological mechanism. In other words, the term IEI was given to people who had a specific cluster of symptoms whose cause could not be proven.[261] As the consensus at the 2004 WHO meeting was that EMF had not been scientifically proven to cause symptoms, the more abstract term *idiopathic environmental intolerance to electromagnetic fields* (IEI-EMF) was approved and adopted. This term persisted for some time in the literature, but lecturers and papers I have listened to and read over the past few years have gone back to the older term, EHS. More recently, in 2025, a group of scientists and those suffering with this condition formed a task force called the One Name Project to come up with a more suitable name, one that would not place the blame of symptoms on the patient. After careful consideration, the task force chose the term *EMR syndrome* to replace EHS.

EHS Symptoms

As the prevalence of wireless technology has dramatically increased, more and more physicians around the world have been studying this phenomenon. Before the COVID pandemic, I attended a medical conference dedicated to EMFs and had the privilege of hearing a lecture by a trauma surgeon from England who decided to dedicate her career to understanding the health effects of EMF. She presented an excellent overview of her clinical experience with EHS patients. During her talk, Dr. Erica Mallery-Blythe reported that electrohypersensitive individuals can present with a myriad of symptoms, most of which are nonspecific (Table 11.2).

• Insomnia/Disturbed sleep	• Tinnitus
• Pressure sense and/or sharp pains in head/ears	• Short-term memory loss
• Headache	• Visual/Hearing disturbance
• Sensations of electric shock	• Tremor/Vibration/Seizure
• Dizziness/Vertigo	• Synesthesia
• Cardiac dysrhythmia/ Palpitations	• Altered energy levels

• Blood pressure anomaly/POTS	• Increased chemical sensitivity
• Musculoskeletal pains	• Increased food sensitivity
• Joint dysfunction	• General sensory upregulation
• Thirst/Dehydration	• Symptoms collection
• Urinary/Bowel urgency	• Circadian rhythm reversal
• Anomia/Thought block	

Table 11.2 EHS Symptoms**
Courtesy of Dr. Erica Mallery-Blythe; EHS Conference, 2021

Though many of the symptoms, like musculoskeletal pain and dizziness, could be caused by many things, some of them—including a vibration sensation, pains in the head and behind the ears, sensations of electric shock, synesthesia, increased chemical sensitivity, general sensory upregulation, symptom collection (meaning the individual accumulates an increasing number of varied symptoms over time), and circadian rhythm reversal—do seem to be more specific to EHS.

A few months after attending this meeting, I met up with a dear friend of mine who was uncharacteristically a little agitated. During our lunch date, she confided that she was nervous because she had been experiencing sharp pains behind her ears. She described these jolts of pain as being unpredictable, and they were making her anxious. She had already been to her doctor, who ordered an MRI of her brain. Her fear was that she had a brain tumor. She knew of my interest in cell phone radiation and asked with remorse, "Do you think it's my cell phone?"

After probing her with a few more questions, such as confirming it was happening behind both ears, I was certain it wasn't a tumor *or* her cell phone. But I did suspect it was from RF radiation. So, I asked her if she used AirPods or another ear-mounted Bluetooth device. She acknowledged that she did use

AirPods for many hours a day. I suggested she put them away for a few months and try to get in the habit of using her cell phone in speaker mode. She seemed a little underwhelmed with my suggestion. But when I checked back with her a few months later, she had surprisingly followed my advice, and her ear pains disappeared. Her brain MRI was normal.

Surveys

Many surveys have been published by epidemiologists and others around the world to assess the prevalence of EHS in the general population. The problem with surveys, though, is that they rely on subjective data that can provide misleading information. I think here in the US, we are well versed in the limitation of surveys. Some communities have lower self-reported rates of EMR sensitivity, such as Sweden at 1.5 percent[262] compared with Austria at 30 percent![263] In 2002, a survey of over two thousand Californians found that 3.2 percent self-reported EMF sensitivity.[264] An excellent study and literature review was performed in 2019 that reviewed surveys performed in various countries, looking at restricted access to work in people with EHS symptoms. By analyzing these studies and more, the authors determined the following:

79 percent of the population experience EMF exposure
symptoms, but are unaware of this association

5 percent to 30 percent of the population have mild
EHS symptoms

1.5 percent to 5 percent have moderate EHS

< 1.5 percent have severe EHS

These authors concluded that 0.65 percent of the general population experience restricted access to work due to EHS.[265] In a world with 8.1 billion people, that translates to 52,650,000 people.

So why isn't this condition recognized by the medical community at large? Chances are that if you go to your doctor and tell them you think you have EHS or IEI-EMF, they won't know what you are talking about! I am only aware of EHS because I have attended conferences during which I was introduced to the condition. Then, I studied it. Furthermore, through my own exploration, I have been fortunate enough to *feel* what it's like to be in an environment in which the only sources of EMF are natural–from the sun, Earth, and other living beings–and contrast that to how I feel in an environment filled with EMF. People with EHS have been described as being "canaries in the coal mine" because they experience a physical reaction the rest of us do not. That should not be reassuring for the rest of us because that doesn't mean we aren't being physically affected by EMF, it may just be that the effects don't rise to a level we can recognize with our bodies. The same is true for most people exposed to virtually any known carcinogen. Just because people don't have immediate health effects from asbestos exposure doesn't mean they won't develop mesothelioma 20-plus years down the road.

Why is it so difficult to prove that it is even a real condition? Part of the problem lies in the lack of standards in the scientific research studying this nebulous phenomenon. As I mentioned, surveys provide subjective data. They ask participants questions like the following:

In the past one month, have you experienced poor sleep? *(Circle one)*

Strongly Disagree Disagree Neutral Agree Strongly Agree

Do you sometimes have ringing in your ears?

Strongly Disagree Disagree Neutral Agree Strongly Agree

Data from this five-point scale is difficult to utilize. Imagine that out of one hundred people surveyed, ten checked off they "strongly agree" they have a hard time sleeping and fifteen checked off they have ringing in their ears. There are dozens of reasons someone might have a hard time sleeping or experience ringing in the ears, one of which may be EMF exposure. Grouping these people together in one category does not make any sense. In addition, what one person describes as a headache can be vastly different from another. People have different pain thresholds. Furthermore, correlating self-described nonspecific symptoms in a diverse group of individuals to a single agent cannot be accepted as proof in the scientific community. There are no lab tests or findings on physical exam that prove someone has a headache, feels dizzy, or endures a subtle sense of anxiety. Finally, there is also no way to prove when a headache or dizziness goes away.

Provocation Studies

Provocation studies have been performed by many researchers as a method to correlate perceived symptoms with EMF exposure. During one form of provocation study, people are exposed to a hidden EMF source, such as a Wi-Fi router not visible to the subject, and then asked if they can tell if the source is turned on or off. These studies have been performed on members of the general population, in addition on those who have been diagnosed with IEI-EMF, to see if those with this condition can "feel" the energy waves or experience a worsening of symptoms when an EMF source is turned on. Other studies have exposed subjects to an EMF source, or a sham as a control, for longer periods of time to see if they develop symptoms. In one study, which was designed as a double-blinded, randomized, controlled provocation study, forty-two subjects from the general population were sent home with mobile EMF exposure-monitoring units and surveys. The subjects were asked to fill out the forms at different times during the day to describe their symptoms, if any. In this way, the researchers were trying to assess whether the subjects were experiencing symptoms at the same time monitors were picking up higher intensities of radiation. These authors concluded that the participants were unable to detect if and when they were being exposed to EMF.[266]

Unfortunately, a study design such as this one provides misleading data. The daily fluctuation of exposure to differing intensities and modulation of numerous overlapping RF-EMF fields in addition to ELF-EMF make it impossible to associate specific exposures to specific symptoms. Furthermore, the effects of EMF are cellular and cumulative. From my own experience, emotional states, such as a subtle sense of anxiety or depression, can take several days of exposure to develop. A transition from the cellular effect of oxidative stress to constitutional symptoms and/or disease might take hours or even days to manifest. The lag time between exposure and disease symptoms may be as long as thirty or forty years in some cases, such as in the development of brain cancer, as described previously.

Proving EMF exposure causes symptoms has become increasingly difficult because most of the population is now *incessantly* exposed to RF and MW radiation 24/7. Finding a control population has become pretty much impossible, particularly in cities where academicians are performing research. Even though it has been nearly impossible to "prove" EHS exists, I am quite certain that it does. I've met too many people, including physicians, electricians, engineers, and others, who have become electrosensitive and developed these symptoms, some even choosing to move to remote regions to try and escape the electropollution most of us live with all the time. Electrically clean environments are becoming increasingly scarce because even remote rural areas are being populated with antennas for broadband access. An extensive web of Wi-Fi networks has been created around the world. Take a look at WIGLE.net to see the dramatic increase in deployment of networks around the world over the twenty years from 2005 to 2025.

Despite the difficulty in finding an electromagnetically clean space on earth to isolate a control population, a research study I recently performed, but have not yet published, evaluated 40 people with ultrasound to see if their red blood cells went into rouleaux formation when exposed to cellphone radiation. The control was the pre-exposure scan. Ten of the subjects studied suffered from EHS. Much to my surprise, *all but one* of the EHS sufferers went into rouleaux. Could the aggregation of blood cells and the impaired oxygen delivery to the brain and heart be contributing to, or be *the* cause of EHS? Seems plausible.

Treating EHS

Until now, due to the inability to prove this condition with reproducible objective data, many people, including physicians, have marginalized the experience of people with EHS. The systemic disregard of these patients is similar to the disregard of those with chronic fatigue syndrome, fibromyalgia, and multiple chemical sensitivity. Because medical professionals don't recognize the cause of their symptoms, patients are unfortunately viewed as psychiatric cases rather than having a true clinical disease. Gratefully, there are physicians now who specialize in environmental disease and recognize symptoms attributed to various conditions, such as Lyme disease and toxin exposures, including mold, EMFs, heavy metals, and others.

The internet has enabled those afflicted with these conditions and others to find support groups and organizations that help people ask questions of those in a similar circumstance, share ideas, learn from one another, and find resources, including physicians who understand and are able to treat these maladies.

Treating EHS may require a complex program and require the guidance of a clinician experienced in this field. Treatment options can involve antioxidants, supplements, and pharmaceutical remedies, but the restoration of health ultimately relies on the ability of the sufferer to remove as much EMF from their environment as possible. Whether or not you have EHS, reducing exposure can benefit everyone and is achieved through a process called *remediation*.

CHAPTER 12

TIPS FOR MEASURING AND REDUCING OUR EXPOSURE TO MICROWAVES AND MILLIMETER WAVES

We are living now in a world filled with staggering amounts of human-generated electromagnetic radiation, especially in the RF range. When I suggest to people that they should lessen their exposure, the first thing I usually hear back is, "I could never give up my cell phone."

We have become used to the accessible information, freedom, and flexibility a smartphone provides. But reducing exposure to MW/RF radiation doesn't mean one has to completely abandon technology. It's not an all-or-nothing decision. There are relatively simple modifications we can make each day to lessen our exposure to RF/MW radiation while still enjoying the use of marvelous technology. Now that you have an understanding of what RF/MW radiation is used for, you will become more conscious of how and when you are being exposed. The more you reduce your exposure, the "cleaner" you feel, and it will likely inspire you to find further ways to clean up electropollution from your environment. I initially learned the theory behind this science at a medical conference, but it wasn't until I began to notice differences in how I *felt* after turning off EMF-emitting devices at home that I truly began to gain an appreciation for the importance of reducing my exposure. Learning how to measure the intensity of radiation being emitted from sources at home is a good place to start this process.

RF Detection

The only way to truly uncover sources of RF/MW radiation in your home and work environment is to use a specialized detector designed to pick up this frequency band. The first detector I owned was about eighteen inches long and looked like a sci-fi weapon. It quickly became a fascination for my daughter. She asked if she could use the "ray gun" to do a science fair project. I was secretly delighted because it also gave me a chance to better define my technique for measuring EMF. I had purchased the detector, but hadn't yet done a systematic evaluation of the house. I had only measured ambient RF levels in each of the bedrooms and emissions from the router and cell phone.

With detector in hand, my ten-year-old daughter went through the house in search of RF/MW radiation. For her, I suspect it was like a treasure hunt. The volume of the sound coming from the device correlated with the intensity of the radiation, and so she found particular pleasure in making the device produce the loudest noise it could. The portable phone system in the kitchen was the strongest emitter after the microwave oven, which was off the charts when it was turned on. Radiation from the microwave oven could be detected throughout the entire first floor of the house, passing through walls, but only while the oven was on. Less powerful emissions came from the Wi-Fi router, the Wii game console, and our cell phones.

In addition to the audio signal, the radiation detector provided a numerical value of the radiation intensity. In order to have consistency, she created a spreadsheet logging the intensity at a six-foot distance from each source and added comments, such as whether or not the noise was continuous or intermittent. She found that the portable phone system's base station constantly emitted strong EMF, regardless of whether or not it was in use. It was an audible indication that the base system was maintaining constant communication with the satellite handsets in other rooms. The Wi-Fi router also emitted constant RF-EMF. Cell phone emissions, on the other hand, were sporadic, at least when the phone was sitting idly on the countertop. When using the cell phone for a voice call, the radiation intensity was much more powerful than we had realized. The experiment was fun and certainly educational, but I had to laugh to myself when I saw the confused expressions of parents and teachers at the science fair as they strolled past her trifold display.

Since then, RF detector design has matured. Now, most units are encased in a sleek plastic container. There are a number of detectors available in the market with a wide variety of prices. In general, as is the case with ELF-EMF detectors, you get what you pay for. If you only want a basic indication of where your exposures are coming from, a less expensive model can be adequate. However, if you are truly looking to know the intensity of your exposure, it is worth the expense to go with a solid company.

Radio frequency detectors typically measure power density, signal strength, or both. The units for these parameters are microwatts per square meter (μW/m^2) and volts per meter (V/m), respectively. Power density represents the average intensity of radiation exposing one square meter per unit time.

The time interval used for this calculation is specific to the detector's software. These meters often provide peak and maximum intensity measurements, which are the most useful because radiation *pulses* have the greatest penetration depth potential. Researchers often use the units µW/cm² and W/m² to express power density which convert to one another as follows:

$$1,000,000 \text{ µW/m}^2 = 100 \text{ µW/cm}^2 = 1 \text{ W/m}^2$$

Detectors that measure signal strength use the unit V/m. This value represents the voltage potential drop over a one-meter distance. If one were to survey a continuous source of RF radiation, such as the power density of a microwave oven, the signal strength in V/m could be calculated as the square root of the product of the power density (expressed in watts per square meter) multiplied by 377 ohms (the impedance of free space):

$$\text{electrical signal strength in V/m} =$$
$$(\text{power density in W/m}^2 \text{ x } 377\,)^{1/2}$$

However, this conversion formula is not accurate for wireless carrier radiation. It's like comparing apples to oranges. But some units display both peak readings in V/m and average intensity values in µW/m².

Coronet, Acoustimeter, and Safe and Sound Pro are three brands I have experience with and recommend, but choosing a detector is a personal choice. Do some research to see which unit best suits your budget and needs. Because my preferred detector is the Safe and Sound Pro II, I will describe this product in greater detail in the following paragraphs. If you choose a different product, the features will likely be the same, or at least comparable.

RF Detector Design

Most handheld detectors available to the public provide a display that includes a numerical value representing the intensity of the radiation detected, LED lights providing a colorized visual representation of intensity, and an audible sound also correlating with intensity. These units do not provide a spectral analysis, meaning a detail of the intensity for each individual frequency the unit is picking up. Rather, these devices provide a number that is a summation of intensities the detector is picking up over the entire frequency bandwidth they are designed to detect–typically between 200 MHz and 8 GHz. If, for example, the detector is picking up emissions from a cell phone and a wireless key fob at the same time, the display will represent the total amount of RF from both sources. By listening to the sound of the detector, with experience, you will often be able to determine the source. A Wi-Fi router, for example, has a specific jackhammer-type sound, whereas a DECT phone produces a *wawawa* sound. Vendors may have resources on their websites to teach you how to differentiate between the different sounds.

Safe and Sound Pro detectors provide three numeric values, each using the unit microwatts per square meter ($\mu W/m^2$).

- MAX (maximum) value represents the highest summation of RF intensity the detector has encountered since it has been turned on, or since the reset button has been pressed (if there is a reset button).
- PEAK value is a continuously changing value representing the maximum intensity the detector has picked up within the previous two seconds.
- AVG (average) value is a continuously changing parameter representing the average intensity within the detectable frequency bandwidth during a two-second time window.

The AVG reading is much lower than the other two values because it, in effect, dilutes peak intensities with gaps of time in between pulsations. This number is perhaps the least useful, although it is the number that industry uses to prove compliance with established safety limits.

A detector's accuracy is an important consideration. Every detector should come with an accuracy assessment, which is measured in decibels (dBs). The Safe and Sound Pro II detector's accuracy is $\pm$ 6 dB on a logarithmic scale, meaning a true intensity value could be between one-fourth and four times the number being displayed. A true measurement for a reading of 1,000 $\mu W/m^2$ could lie anywhere between 250 and 4,000 $\mu W/m^2$. Although this might seem like a wide range and therefore not useful, in the world of RF detection, this is considered excellent accuracy. This detector can measure radiation intensity between 0.005 and 2,500,000 $\mu W/m^2$. Therefore, if the detector displays 2,500,000 $\mu W/m^2$, the maximum level the detector can measure, the true intensity of exposure is likely higher.

Surveying Technique

These detectors often use a bidirectional antenna. This means the orientation of the antenna with respect to the radiation source will affect the measurement. It is therefore necessary to systematically change the orientation of the detector as you measure to ensure frequencies from *all* directions are accounted for. If you hold the detector in an outstretched hand and maneuver it in a figure-eight pattern across the body and slowly rotate your body in a circular fashion, you will detect radiation from all directions.

Consider Distance

Just as with surveying ELF-EMF, RF radiation intensity diminishes significantly with increasing distance. When you measure a radiation source, take into account how far away from the emitter you are when taking a reading. Try and take measurements at meaningful distances from your devices. For example, if you're assessing the RF emitting from a Bluetooth headset, you would need to survey the intensity right next to the headset rather than three feet away because when it is placed in or on your ear the distance between the emitter and your brain is only a couple of inches. Similarly, if your Wi-Fi router is only a few feet from where you sit at your desk, measure the RF intensity of the router from that distance.

As you did with ELF and magnetic field strength measurements, create a spreadsheet and chart places in your home and office where you survey (see appendix D for a sample spreadsheet to record RF intensities). Record the MAX intensity in the center of each room; the PEAK and AVG values are less useful. Take measurements of specific devices, like your cell phone, wireless speakers, router, portable phone, baby monitors, cordless mice, cordless keyboards, security system, smart home appliances, etc., and add them to the chart along with the distance from which you measured each. Survey your office, front walk, and backyard to get a sense of the overall environment you are living in. Once you begin surveying, you will likely become aware of additional sources of EMF in your home. It can be hard to wrap your head around just how much exposure we are all receiving every day and night. Understand that the "background" level for RF radiation is 0.0. When you are finished, study your results. Determine where your major exposures are originating from.

Detection of higher 5G frequencies in the millimeter range require a different meter, one that can detect frequencies over 8 GHz. Millimeter wave detectors have come to market. Safelivingtechnologies.org sells a 5G meter called the Safe and Sound Pro mmWave meter that can detect MMW between 20 GHz and 40 GHz, a bandwidth that encompasses the high-band frequencies currently being emitted by the mobile telecommunications industry antennas for 5G networks. Note that if you want to survey for *all* 5G-related frequencies, an additional detector, such as the Safe and Sound Pro II, will also be needed to detect the low- and mid-frequency bands employed by the 5G network.

The Building Biologist

If you are feeling overwhelmed at this point, consider hiring a professional to come into your home and do a survey for you. A building biologist will typically perform a whole-house survey for magnetic fields and ELF-EMF, VLF, LF, RF-EMF, and MMW radiation. These professionals have much more sophisticated equipment that can provide a spectral analysis of each frequency detected and often determine its source. A good building biologist will offer suggestions for remediation. I hired one of these professionals after I moved into the city. Confronted with so many sources of exposure, I didn't know how to begin the remediation process. The Building Biology Institute website (build ingbiologyinstitute.org/find-an-expert/) is a great resource for obtaining a building biologist. Check them out. Members are located all over the country and can test for all forms of EMF in addition to other indoor pollutants, including mold.

Hardwiring and the Removal of Extraneous Sources of Radiation

After you uncover the sources of RF radiation in your home, the next project will be to reduce excessive and eliminate unnecessary exposure. We have all been trained to turn the lights off when we leave a room. Why leave a device on when you're not using it, particularly when the cost of electricity has gone through the roof? I cannot provide you with a specific safe level of intensity for your home. Background is 0.0; anything over that is electropollution. Realistically, it will be nearly impossible to completely eliminate RF/MW radiation from your home, but I suggest you try to lower your exposure as much as you can, paying particular attention to areas of prolonged exposure, and especially your sleeping areas.

The Big Three

After my daughter's survey, we initially tackled the three largest sources of emission: the portable phone base station, the microwave oven, and the Wi-Fi router. The portable phone system was permanently unplugged and replaced by a corded land line. The base unit for the multiunit DECT phone system had been right next to the kitchen sink. A few days after disconnecting this system, I felt a familiar sense of calm while in the kitchen. Over the years, I had grown accustomed to a subtle angst while doing dishes, as if I were being hurried. It was a sensation I wasn't conscious of, but once the phone was removed, that feeling disappeared.

The next biggest emitter was the microwave oven. Although I never liked heating up food in the microwave, I did use it on occasion to heat up tepid coffee. But after the science fair experiment, I unplugged the microwave and had it permanently removed from the kitchen. I know people who think they can't live without a microwave. But not once have I wished I still had a microwave oven. Food tastes much better when it is cooked in a regular oven, toaster oven, or a slow cooker. And, although vendors claim microwave ovens don't leak radiation, that is not now, nor has it ever been, true. When you do your home survey, you will no doubt find your microwave oven leaks radiation as well. I have surveyed many microwave ovens, some even while being interviewed on TV news shows, and they always leak *a lot* of radiation. Even standing at a six-foot distance from the oven, I have recorded exposures over the meter's maximum intensity limit of 2,500,000 µW/m².

The third major emitter in our home was the Wi-Fi router. Eliminating Wi-Fi from our home was more challenging. Like most of you, we relied on Wi-Fi. In an initial attempt to limit our exposure to RF radiation from Wi-Fi, I placed the router on an appliance timer, which turned the unit off a half hour before bedtime. The goal was to prevent the disruption of circadian rhythms. However, while homeschooling during the pandemic, my daughter began to complain of headaches and, suspecting the headaches were coming from RF/MW radiation, I had ethernet data ports installed throughout the house. The Wi-Fi and hotspot capabilities of the router were then turned off by our internet provider, and from then on, we only used an accessory router for Wi-Fi access on rare occasion. I bought us all specialized adapters for tablets, laptops, and cellular phones so we could turn the Wi-Fi antennas on our devices off while in the house. In addition, I filled a basket with various adapters for houseguests so they too could also get online via a hardwired connection. The kids were initially upset by the hassle, but they slept well and maintained normal personalities and mood during that trying time. In retrospect, it was a terrific decision and one I highly recommend everyone make. I have lived in a home without Wi-Fi for over three years, and I don't miss it.

Reducing EMF in the home is manageable, but will require some adjustment. To some, this may seem like going back in time ten-plus years. Technology is rapidly advancing toward wireless automation and online control of nearly everything in your home. But I recommend stepping out of the march and going old school. Try and hardwire everything you can. If you still have a landline, make it an old-fashioned corded phone. If you want multiple handsets, have an electrician install phone jacks for additional corded phones around your home. This is simple to do and how we lived before cordless phones became mainstream. If you are so inclined, you can even have your cell phone calls forwarded to the landline while you are at home and then turn your cell phone off without needing to worry about missing a call.

Removing Sources of RF/MW from the Bedroom

After tackling the most intense emitters in the home, it is crucial to try and sleep in an electromagnetically clean bedroom. Ideally, the intensity of RF/MW radiation in the environment where you sleep should be < 3 $\mu W/m^2$. Depending on where you live, this may be difficult or even impossible to achieve without the use of a sleep canopy, which will be described in detail under "Shielding Fabrics" (see page 251). Many people have a Wi-Fi router in their bedroom and don't realize that's probably why they need to pop an Ambien to sleep. If this is the case in your home, please unplug the router and move it as far away from your bedroom as possible. If this is impossible, unplug the Wi-Fi router before you head to bed. If you need an electrician to install a new data access port, do it. If you have a Wi-Fi extender or booster in your bedroom, remove it. If you have a smart television in the bedroom, it can be hardwired with an ethernet cable to a data port, then go into the settings and turn off the antennas. Measure to be sure you got them all.

I highly recommend you get into the habit of turning your devices off or placing them in airplane mode for the entire night while you sleep, and for a portion of your waking hours. You will feel *so* much better! My suspicion is you won't miss out on too much. Many people use the clock function on their cell phone as an alarm clock. This function will work even if the phone is in airplane mode. If you sleep with a cell phone underneath your pillow, try to break that habit. If you are someone who must have your cell phone turned on and active in your bedroom at night (an emergency first responder, for instance), keep it as far away from your bed as possible or sleep in a shielding canopy and place the phone outside of it. In addition, be sure to turn off all unnecessary antennas, including 5G, Bluetooth, Wi-Fi, hotspot, locator, etc. This is easy to do, once you know how.

For Apple products, the list of antennas that can be manually turned off can be accessed under the settings tab. When the antenna name says *Off*, it is off. If it reads *Not connected*, the antenna is on and emitting radiation in search of a connection. Note that in the home screen, if you swipe down from the top of the screen to manually turn off antennas on the utility page that opens, up, antennas may remain on even if you think you are turning them off. It is best to go and turn off antennas under the settings tab. Also note that your phone can be in airplane mode and have the Wi-Fi antenna on.

Do you sleep with any devices on your body? Some folks wear wireless products that collect data while they sleep, such as how many minutes are spent in REM (rapid eye movement) sleep or their heart rate variability index. I have no idea how accurate these devices are, but I do know that if they operate with Bluetooth technology, they are exposing you to unnecessary RF-EMF while you are sleeping. I suggest you remove them and turn them off. If you can't turn them off, place them in a room far away from your bedroom while you sleep. The same is true for smartwatches, smart rings, or anything else you might like to wear during the day to keep track of things. When asleep, you should be unplugged.

If you have a medical device that must be connected to Bluetooth, there may be some hardwired options. Some sleep apnea machines transmit data by emitting RF signals all night long. If these units transmitted data once a week during the middle of the afternoon, it would be no big deal. But instead, they pulse RF nearly constantly. It may require being a little emphatic, but companies that make sleep apnea machines do offer units in which you can record data on a card instead of having the data transmitted wirelessly. At the end of the month, you will need to mail the company your data chip to document compliance. Be aware that sleep apnea devices also typically produce strong magnetic fields, so keep them at least two to three feet away from you while you are sleeping.

Wireless pacemakers and implantable loop recorders that record information from your heart are becoming commonplace as have wireless hearing aids that connect to Bluetooth. If your physician recommends one of these products, the benefit of having the device will likely greatly outweigh any potential small risk from RF radiation, but you could ask your provider if there are any options that don't emit constant RF. They may look at you quizzically for now, but I suspect physicians will become more cognizant of the RF radiation produced by these products in the future and get on board with the need to go hardwired within a few years. Maybe then, the physician community will convince manufacturers to make safer products. Remember that physicians used to smoke cigarettes. Sometimes it takes a while . . .

Hardwiring the Home Office

The home office is another common EMF "hot spot." The wireless mouse, wireless keyboard, wireless printer, and desktop computer all emit RF-EMF and can be hardwired. Converting over may mean buying new peripherals, but wired varieties are usually less expensive. I used to find sitting at my desk to be somewhat stressful. I suppose it was in part because I was often paying bills and it meant money was going out instead of coming in, but it was a feeling similar to the angst I experienced while washing dishes. After removing all of the wireless peripherals in my office and plugging into an ethernet cable for internet access, sitting at my desk is now *usually* more relaxing. Be aware that many devices have more than one antenna, and it is important to survey with your meter to ensure you have turned off *all* antennas before taking comfort in your newly hardwired office.

The Living Room and Other Common Areas

Depending on the number of people in your household, the living room may be a more difficult space to clean up. In my home, I asked the electrician to install three data ports to enable two people to access the internet at the same time. A smart TV plugs into the third port with an ethernet cable. As you can imagine, I am not a fan of Alexa, Siri, or any other voice-activated "home helpers." They constantly beam out RF/WM radiation to keep in touch with all the items in the house they control. For the same reason, I recommend opting out of smart lights and smart appliances. When purchasing a new washer and dryer, I was recently confronted with a selection of appliances that connected to the internet of things (IoT). Just what I want: a washing machine that beams up signals to whomever might want access to my personal laundry habits. No thanks. All smart appliances emit RF-EMF constantly and will expose you and your family to significantly more ambient radiation at home. A friend measured her new smart stove, and it emitted RF radiation with a power density of 1,000,090 $\mu W/m^2$!

Smart Meters

Digital meters recently installed by utility companies allow them to access usage data without needing to send an employee to each home every other month to check meter readings. Not all digital meters are smart meters. But if your digital meter lists an FCC ID# on its display, you have a smart meter. There was a time when you could opt out of a smart meter, but now, in most instances, utilities can install them without your consent. Companies may put a meter on your home, and you may not even know it. I have heard people describe how they are no longer able to sleep or have developed new anxiety after a smart meter was installed on their home.

Smart meters provide the company with near real-time usage data. When are you home? When are your appliances turned on? When do you use water? Even without considering the radiation exposure, the surveillance alone is creepy. This technology obviously saves utility companies lots of money, as they don't need to employ as many technicians. But these devices only *need* to report back to the mother ship once a month to be adequate for billing purposes. Even once a week would be inconsequential. But these meters transmit data nearly *constantly*. Why does the utility company need this kind of data? Every time the meter transmits data, it does so by producing a strong pulsation of RF/MW radiation that can easily penetrate into the home. These meters also generate strong magnetic fields and create continuous currents of dirty electricity.

Remediation of a smart meter requires dealing with both dirty electricity and RF/MW radiation. I've heard of some people convincing the utility company to move the meter off the house and onto a post closer to the street. But if this is not an option, there are small circular metal Faraday cages that have been designed to fit over the smart meter, which can be found online. They don't work too well, but they might help a bit. Another option is to use a shielding paint on the interior of the common wall the meter has been placed on to reflect the majority of RF radiation back to the outside. In order to combat the dirty electricity created by these meters, I recommend a whole-house dirty electricity filter, discussed in chapter 5.

Making moves to limit RF/MW emissions in your home, and particularly in your common areas, won't go unnoticed. People often enter my home and without prompting express how they feel relaxed and comfortable. Facial expressions sometimes communicate a subtle confusion, as if they don't really understand *what* is different about my home. Is it the decor? The air? The artwork? When people tell me they feel relaxed in my home, I usually let them know there is no Wi-Fi and this often leads to a conversation about the importance of removing RF-emitting devices from the home.

Shielding

Once you have eliminated as many sources of RF-EMF from inside your home as you are comfortable with, resurvey and see how intensity values have changed. Are you done? Or, is there still a significant amount of RF-EMF in your home? If there is, are you aware of where it is coming from? It may be that you are receiving RF radiation from outside your own four walls. If this is the case, you should consider shielding.

There are many different shielding materials available, including fabrics, screens, window films, paints, and tape. Usually, a combination of materials can be used to reduce the ambient levels in a room. If you research shielding products, you'll find their effectiveness is described in terms of decibels (dB). Decibels are a unit of comparative intensity that employ a logarithmic scale, similar to sound. The formula for determining effectiveness is:

$$x \text{ dB} = 10 \times \log_{10} (\text{original intensity} \div \text{new intensity})$$

If you surveyed an area with an initial intensity of 100,000 $\mu W/m^2$ and a shielding product, such as a window film, reduced exposure to 10 $\mu W/m^2$, the effectiveness of the product in decibels could be calculated as follows:

$$100{,}000 \ \mu W/m^2 \div 10 \ \mu W/m^2 = 10{,}000$$
$$10 \times \log_{10} 10{,}000 = 40 \text{ dB}$$

The window film in this case would have an *attenuation rating* of 40 dB.

If shielding paint that reduces intensity from 100,000 $\mu W/m^2$ to 1 $\mu W/m^2$ is applied to a wall, the calculation of dB would be as follows:

$$100{,}000 \ \mu W/m^2 \div 1 \ \mu W/m^2 = 100{,}000$$
$$10 \times \log_{10} 100{,}000 = 50 \text{ dB}$$

Shielding Fabrics

Shielding fabrics are particularly easy to work with. These special fabrics are threaded with metals, typically silver and copper, and are highly effective at reducing RF exposure levels when used properly. EMF-shielding clothing is available, but its value is probably dependent on the style of clothing and how well the fabric is maintained. If one wears an article of EMF-shielding clothing over the upper body, exposure levels will be diminished, but there will still be exposure to uncovered parts of the body. Because EMF refracts and reflects, it will wrap around the shielding fabric margins and penetrate parts of the body beneath the fabric, potentially then reflecting off the undersurface of the material. The larger the piece of the material and the more body area it covers, the more effective a shielding fabric will be. A shielding tank top seems pointless, whereas a hoodie may be more effective. I have no personal experience with EMF-shielding clothing, but if you anticipate being in an environment with high RF exposure for a prolonged period of time–such as a long airplane trip, a movie theater, or a concert–and you feel better wearing EMF-shielding fabrics, then go for it!

The most common use for shielding fabric is to create a sleep canopy, which typically consists of four side panels and a top. The side panels wrap around dowels suspended from the ceiling, creating a tent. If your bedroom is on a ground floor, and there is no radiation source below your bedroom, you may not need a ground sheet made of the same fabric. But if you live on a second or third floor, you will most likely need to place your bed on top of an additional sheet of shielding fabric so you effectively create a Faraday cage. Light is able to penetrate through the material, but RF radiation and bugs cannot. A sleep canopy will dramatically reduce your EMF exposure while sleeping. Outside of the canopy, RF radiation levels may be over 20,000 µW/m², but inside, you should expect values less than 10 µW/m², and ideally less than 3 µW/m².

I have been sleeping in an RF-EMF-shielding bed canopy for four years. When I first moved into the city, I found I couldn't sleep. Months went by with me tossing and turning in bed. The router was turned off and all cell phones in the house were in airplane mode, yet I couldn't seem to relax. I hired a building biologist, who surveyed my home and uncovered external sources of RF that were coming into my home from the neighbors' networks and outdoor 5G/4G LTE antennas. The biologist recommended I purchase a sleep canopy because he felt it would be difficult to remove all the ambient RF from my bedroom.

On the first night after the canopy installation, I finally had a good night's sleep. It was a godsend to me, as I had been exhausted from months of insomnia. I had a few excellent nights of sleep, but then, one night, I again didn't sleep well. I lay in bed filled with despair, thinking I had experienced some sort of placebo effect that had worn off. When the room filled with morning light, I opened my eyes to see a flap of the canopy had been folded back. I smiled to myself, realizing the gap in the fabric allowed EMF to enter the sleeping space. Like light from a hallway coming into your bedroom through a crack in the door, it must have been enough exposure to interrupt my sleep. That day, I went to the store and purchased fabric clips, which I have subsequently used each night to ensure the canopy edges overlap and don't fold back again during the night.

Although I am hesitant to make the creation of a sleep sanctuary more complicated, I have discovered that sleep canopies do not effectively block the high-frequency band of the 5G network. So, if you are stuck with a nearby small cell antenna outside showering your home with high-band frequencies, you will need to take an additional step to shield your sleeping quarters. This may take the form of shielding paint or even putting up a metal shielding product, such as aluminum foil, between the bed and the antenna. This type of shielding will need to be done in conjunction with a MMW detector. Every situation is different, and in some instances, a building biologist may be very helpful.

Other Shielding Materials

Some paints, foils, films, and fabrics are highly effective and rate over 98 dB! All products have their own unique formulations and, as long as you understand how they rate, you can pick the right ones for your needs. The only hassle and potential dilemma will be gaps in shielding that will allow energy to penetrate through your barrier. As mentioned earlier, you can have a whole bedroom painted with shielding paint and reflective film on the windows, but if there is a gap underneath the door to the hallway and there is a router on in the hallway, RF/MW radiation will leak into your room from the Wi-Fi router just as light from a lamp would.

When I first started talking to people about the potential health ramifications of all the RF radiation in our environment, I often found myself the butt of "tinfoil hat" jokes. Interestingly, now that there are 5G antennas dotting the landscape, I haven't heard a tinfoil hat joke in years.

One of my first RF shielding attempts did actually involve aluminum foil. I didn't wear the foil as a hat, but I did attempt to use it as a shielding material by draping a few sheets of aluminum foil over our Wi-Fi router antennas. It did decrease the RF intensity in the house, without limiting our ability to use the network. Over time though, burn marks formed on the undersurface of the aluminum and, soon after, actual holes formed where aluminum had been burned away by the radiation. Concerned about breathing in aerosolized aluminum and the possibility of a fire hazard, I removed the foil tent and opted to plug the Wi-Fi router into an appliance timer instead.

Shielding Precautions

Shielding products are conductive. It is therefore important to ground these materials. Although vendors in this niche often sell grounding kits, I recommend having an electrician perform this to ensure grounding is done properly.

Shielding products reflect RF/MW radiation. Therefore, once an electromagnetically clean environment has been created with shielding material, do not bring wireless devices inside. The device will receive low levels of signal and therefore produce more intense RF in an attempt to communicate with the network. The RF/MW radiation produced by the device will reflect off the walls and other shielded surfaces, increasing your exposure.

I have heard some raise the question of a potential downside to shielding. They fear shielding can remove you from the natural frequencies of the Earth and disrupt your normal physiology. I believe that theory is based on research by R. Wever, who in 1970 published a paper showing that volunteer subjects who lived in an underground apartment shielded from electric and magnetic fields demonstrated circadian rhythm disruption–a finding that did not occur in a control group who also lived underground but in an apartment without shielding.[267] I'm not sure circadian rhythm disruption would occur in someone who is spending only a portion of their day or night in a shielded environment. Furthermore, shielding products operate within a specific frequency bandwidth. Fabrics and films are designed to block EMR in the microwave range, not the extremely low frequency of the Earth's Schumann resonance. With my current understanding, I don't think shielding products should cause a health problem.

Personal Devices

The cell phone is the single most powerful source of daily RF/MW exposure for most people. Whether or not you spend your time talking on the phone, texting, or carrying the phone around waiting for notifications, cell phones emit frequent bursts of radiation, searching for updates and messages, and relaying location data to apps.

It is relatively easy to decrease the intensity of your exposure to cell phone radiation. Smartphones give you the option to turn off antennas. If you are not currently using a Bluetooth device, turn off the Bluetooth antenna. Similarly, if you are not using Wi-Fi function on your cell phone, there is no reason to have the Wi-Fi antenna on. If you turn off the Bluetooth and Wi-Fi antennas, in addition to the 5G, and hotspot antennas, you will substantially decrease the emissions from your phone. As mentioned previously, be advised that if your phone says, "not connected," it is still emitting radiation in search of a connection. The Wi-Fi and Bluetooth need to be turned off individually to cease this unnecessary radiation exposure. Another option is to use a Wi-Fi connection to make a call with the Wi-Fi calling feature many cell phones now offer. On an Apple phone, this option is listed under the cellular data menu. If you use this option in public places and at home (if you keep Wi-Fi on during waking hours), you will lower your radiation exposure from 0.6 to 3.0 watts to a maximum of 600 mW, reducing your exposure significantly.[268]

Additional behaviors that are relatively easy to change include carrying your phone in a bag and avoiding carrying an active cell phone in your pocket–and certainly not in a bra or waistband! If you need to place it in your pocket, put it in airplane mode.

Additionally, I recommend you never hold a cell phone up to your head. Use the speaker mode function or plug in a headset if you want privacy. Despite their incredible popularity, I do not recommend syncing up with Bluetooth earbuds. Even though the of RF radiation from a Bluetooth set is much less intense than from a cell phone, the emission is directed right into the ear canal.

"Wearables," including smartwatches and personal fitness monitors, also produce significant RF radiation. According to my detector, an Apple smartwatch emits a power density up to 54,000 $\mu W/m^2$ at the skin surface and 860 $\mu W/m^2$ two feet away, the typical distance to one's head from an outstretched arm. A Fitbit fitness tracker produces blips of RF radiation every second with power densities up to 70,000 $\mu W/m^2$ at the skin surface and 50 $\mu W/m^2$ at a two-foot distance. Although both devices are worn on the wrist, consider that blood vessels in the wrist are superficial and only a few millimeters deep to the skin surface. Red blood cells, immune cells, nerve cells, muscle cells, and all the cellular components of the skin are exposed to this radiation when one of these devices is worn. Considering it typically takes blood cells less than one minute to circulate through the entire body, if one applies a wearable device on the wrist for an hour or more, a large percentage of your blood cells will be exposed to its radiation over that duration. As you can imagine, I am not a fan of these devices.

Tablets can be hardwired and connected to an ethernet cord by using an adapter. Do a search online to see which adapter is needed for your individual unit, as they are vendor and model specific. Once you are hardwired and plugged in, you can turn off the Bluetooth and Wi-Fi antennas on the laptop and surf the web without any exposure.

Crystals, Shields, and EMF Counteractive Devices

Before we understood how EMF damages cells, the physical mechanism behind it, a number of products came on to the market that offered symptomatic relief for those suffering from EMR syndrome (EHS). These included crystals, shields, and other devices.

Crystals

Many believe that crystals–in particular shungite and/or orgonite–can counteract the negative health effects of EMF. Orgonite is a man-made composite of resin and metal shavings that can be molded into any shape and is frequently sold in pyramidal form or as a pendant. Shungite, on the other hand, is a naturally occurring mineral from Russia that is mostly composed of carbon atoms. Although many dismiss these beneficial claims as unsubstantiated rubbish, I am not so quick to do so. I believe it is complicated and that there are probably two major categorical effects that seem to be at work when one holds an RF-emitting gadget such as a cell phone.

First of all, I have witnessed a technique called "strength testing," also referred to as "muscle testing," demonstrate muscular weakness occurring immediately after someone picks up an active cell phone and disappearing when they put it down. This is a technique often employed by chiropractors and other complementary and alternative healthcare practitioners. I've performed this technique myself on many people, and the weakening effect is consistent and reproducible. It may be that the RF radiation from the cell phone causes interference, creating noise that disrupts the coherence of the body's innate electromagnetic field. This weakness is improved when the phone is put down, but can also be eliminated when the person being tested holds a shungite crystal in the opposite hand. I have also witnessed this restorative effect when one holds a static magnet from the company Nikken, described in chapter 15, in the opposite hand. The weakening of the body's energy field by a cell phone is an intangible effect, but coherence is an important state of being that I believe may help one avoid disease. With that being said, no scientific experiments have documented this effect.

This is not the whole picture though. In addition to electromagnetic interference, EMF causes oxidative stress, which can then result in a number of health repercussions previously discussed. Crystals that improve systemic incoherence are likely not going to affect the prevalence of oxidative stress. For that reason, I suspect that these amulets have value, but beware of a false sense of security that you are "protected" from the harmful effects of EMF. This viewpoint is shared by others. An excellent overview of shielding methods and products designed to protect against EMF was published by the scientist Dimitris Panagopoulos in 2019.[269]

EMF Shields

There are pendants, stick-on materials, and other devices that claim to diffuse EMF and make them less harmful. I have tried out a few of them, but assessment of their effectiveness is difficult. If you place a stick-on EMF shield on a device and it is still able to receive and send out a signal, it is still emitting RF radiation. There is one company called Aires Tech that has performed many experiments on their RF shielding products, which are available for review on their website. The research seems valid, although the peer-reviewed literature is sparse. They have looked at changes in HRV among other potential biomarkers. I have not read any peer-reviewed literature confirming the clinical effectiveness of these devices, but I imagine it would be hard to get this kind of study published.

Phone case and tablet shields may be helpful but need to be used with caution. These materials reflect EMF and can therefore *amplify* the signal in front of the shield. These sleeves may block EMFs, but be aware that a device will produce a larger power density if it is having a difficult time communicating with the network. I place my cell phone in airplane mode anytime I need to put it in my pocket, and that has worked out fine for me.

EMF-Emitting Counteracting Devices

There is a theory that human-generated sources of EMF are particularly damaging because they are erratic and noisy. Think of the difference between hearing a note played on an instrument and a noise such as static. Noise creates stress, whereas the tone creates calm and, in some cases, resonance. Naturally occurring sources of EMF, such as that from the Earth's magnetic field or sunlight, are in a sense, tonal. They have a predictable repetitive wave form and constant frequency.

When humans create RF radiation to disseminate information through wireless transmission, modulation of the waveform creates an erratic and irregular signal. Some believe the erratic nature of the waveform causes anxiety and incoherence in the human energy field. EMF-emitting devices that produce a predictable, more natural bland source of EMF have been touted as normalizing the human energy field and counteracting the random frequencies associated with WCR. After experimenting with this technology for a year, I believe that, similar to crystals, the overriding effect of the superimposed signal may have some value, but probably does not protect one from the harmful effects of oxidative stress. If you enjoy the ambiance created from these emitters, I don't see a downside. A device such as this may make your energetic coherence better, but it is best not to let these devices convince you it is OK to have a smart home or to disregard other methods to reduce RF/MW radiation exposure.

In conclusion, once you have identified sources of RF/MW radiation, there are many options available to reduce your radiation exposure levels. Knowing provides power and choice. Turn off devices and remediate as much as possible. Then, feel free to use EMF-blocking crystals, shields, pendants, and signal-counteracting devices as if they were frosting on a cake. People perceive different benefits from these products just as they experience different symptoms from EMFs.

Although muscle testing has been disparaged as pseudoscience by some, it has been shown to have value by others.[270] This technique can take many different forms and can be performed on oneself. A quick online search will provide you with many different techniques to learn this process. Feel free to self-monitor your response to various products to see if you perceive a favorable response.

CHAPTER 13

THE BENEFICIAL USE OF EMFs IN HEALTHCARE

Although it seems to be true that incessant exposure to man-made electromagnetic energy fields can be harmful to health, we depend on natural energy fields for our existence. Cells within a life-form communicate through low-frequency fields, and unspoken communication among organisms also occurs via various frequencies of electromagnetic radiation, although this science has not yet been fully explored. Natural sources of radio frequency and microwave radiation, on the other hand, do not exist on Earth in any significant amount.

Naturally occurring magnetic fields create an energetic framework in which living systems navigate. How else could a cell or an organism really know how they are oriented in space, or in which direction they are travelling? Many animals don't have a visual system. The Earth's magnetic field provides orientation. Even though we, as human beings, might think we don't use the Earth's magnetic field to function, we do. Some scientists postulate that tiny hairs in the inner ear of humans and other animals respond to the Earth's magnetic field, providing us with an important sense called *magnetoreception*.[271]

Our bodies and our individual cells need to know where they are in relation to one other in space. Organ systems need to grow in a certain shape and a specific location, and for the most part, there is incredible consistency.

We are beginning to understand that biological systems are dependent on ELF-EMF, and that the wrong frequency or intensity of radiation can do harm. But can the right frequencies and intensity of ELF-EMF prevent us from getting sick or even heal us if we become injured or infirmed?

Research is ongoing, but it is a slow process. The present mainstream healthcare delivery system is based on the premise that there is an anatomical and a biochemical basis for disease. Surgeons can correct anatomical problems, and pharmaceuticals are supposed to address biochemical imbalances. Medical doctors are taught that if your patient has a disease or disorder, there is a pill that can help. Antibiotics, antivirals, vaccines, chemotherapy, immunotherapy, steroids, nonsteroidal anti-inflammatories, insulin, blood pressure medications, etc. are the arsenal with which allopathic medical doctors treat disease. By limiting treatment to surgical and biochemical interventions, conventional allopathic medicine, as I learned early on in my career, is *very* limited. Although the knowledge base doubles every seven years, the greater understanding, for the most part, goes deeper into the already accepted paradigms of drugs and surgery.

Even the understanding that there is a foundational biochemistry to life is relatively recent. I remember my father telling me a story of how when he was eating dinner one night at college in the early 1950s, they announced that Watson and Crick had discovered DNA. It's hard to fathom the advances in our understanding of biochemistry and genetics since that time. But genetics and the use of drugs to treat disease are based on the assumption that biochemistry and anatomy are the foundational blocks of life and therefore need to be corrected to treat disease. This is a mechanistic view of life and medicine. But evidence abounds that biochemistry is *not* the foundational pillar of life. Energy underlies biochemistry and creates a framework for anatomical structure. In his book *The Body Electric*, Robert Becker detailed his experiments in which he showed that manipulation of energy is behind the regeneration of limbs.

Grounding

Humans often live in buildings on padded, insulated floors, and walk outdoors in footwear with rubber soles. As a result, everyday actions, like walking on a carpet, can result in the transfer of electrons from a material to the body. The accumulation of electrons, referred to as static electricity, will eventually discharge as an electric shock when the opportunity presents itself, such as when one goes to flip a light switch.

On the other hand, when a body is grounded–meaning it is literally in contact with the Earth–excess electrons immediately discharge through our feet or whatever body part is in contact with the soil preventing the buildup of static electricity. Wet soil will transmit current more readily than dry soil.

Most animals are always grounded, at least when they are walking outdoors. Some of us, however, are not grounded at all during the entire day. We have become, in a sense, isolated circuits. Research has shown it is better for one's health and immune system to discharge static electricity through grounding.[272] When an animal injures itself, the injury can and usually does heal miraculously quickly. The reason may lie in grounding.

The cardiologist Stephen Sinatra was one of the pioneers who studied the health benefits of grounding and popularized the concept of *Earthing*. It is simple. Allow your skin to come into contact with ground as much as possible. Because of the body's conductive nature, contact between a finger, an elbow, a foot, or a toe and the Earth–or any conductive material grounded to the Earth–is all you need to do for your whole body to be grounded. By contacting the soil, the leaves of a plant or tree, natural water sources, and rocks, we become grounded, and pent-up charge immediately dissipates back into the earth. One of the physiological effects of this release is relaxation and stress reduction.[273] Stress reduction can translate into overall improved health. Some forms of footwear will allow you to ground, like purely leather-soled shoes. But plastic or metal layers in a foot bed construction will break the circuit between your body and ground. Several companies sell grounding shoes, including Harmony 783.

Dr. Sinatra has now passed, but his company sells products that you can plug into a properly grounded outlet in your home to enable you to ground even if you are lying in bed or sitting at a desk. You may not feel anything different when you put your feet on a grounding pad, but grounding does provide a subtle sense of relief that, given repeated exposure, you may learn to recognize and find relaxing during the day. If you do try these products, be sure to use their testing device to make sure your ground is properly grounded. In addition, make sure you don't have any dirty electricity contaminating your ground, for if you do, you may inadvertently expose yourself to unwanted stray voltage. Dr. Laura Koniver's book *The Earth Prescription* provides dozens of creative ways to ground naturally during each season. It's a helpful resource. I try to ground as much as possible and recommend this important practice, particularly given our constant exposure to RF radiation and EMFs.

Healing with Magnetic Fields

It seems most physicians who veer off the conventional path of medicine have a story. We have an event in our life that teaches us that–despite the vast knowledge base accumulated by academic physicians and scientists in the world–there are still limitations and significant gaps in our understanding of how living systems get hurt or heal. I am certainly no exception. My first unexplainable encounter with a healing force beyond the disciplines of the medical school toolbox involved static magnetic fields.

The first time I was introduced to the healing power of magnets, I was freshly out of a radiology residency program, working a private practice job with two other physicians at a local hospital. This was a transition year, during which I had planned to spend a few months in East Africa teaching radiology to doctors at the Kilimanjaro Christian Medical Center in Moshi, Tanzania. I had planned a trek up Mount Kilimanjaro, the tallest mountain in Africa, among other sightseeing activities.

Two months prior to the scheduled trip, I took a vacation to Mount Rainier in Washington State with my cousin. I had been training for Kilimanjaro and thought hiking Mount Rainier would be a good practice run. On the way down the slope, I stepped off a big rock in an awkward manner and felt a squish in my left knee. *Uh oh*, I thought. It didn't take long for the knee to swell up and become stiff and painful. By the time we got to the cabin, I was humiliated with my arms draped over my cousin's shoulder.

That evening, we rested in a hot tub and discussed the day's events. My cousin had spent many years in England and approached healthcare in a totally different manner than I did. She mentioned that she had come prepared with arnica, a compound I was unfamiliar with at the time. After we got dried off, I skeptically applied the ointment to my knee.

Upon my return home, I promptly saw an orthopedic surgeon who examined my knee. He mentioned arthroscopy. I told him of my plans to go to Africa, so he agreed to try a conservative regimen of anti-inflammatories to see how things unfolded. Over the next week, the knee pain slowly improved.

As expected, the swelling went down and I was soon able to walk without pain, but there was a little annoying sensation that persisted with each step. At the time, I drove a car with a manual transmission, and each time I pressed down the clutch, I felt a twinge in the knee. Each subtle jab was disheartening. The time for my trip was fast approaching. After the prescription ran out, the knee pain came back with a vengeance. I was despondent and considered my options, including canceling the trip. But I had already paid a hefty sum to not only trek Kilimanjaro, but to also go on safari and go gorilla trekking in Uganda. It was to be a trip of a lifetime, and I wasn't willing to give it up without a fight.

One of my coworkers had been watching me limp and listening to me complain about my knee for a few weeks. In a casual manner, he asked if I was willing to try an alternative therapy. "Do you have an open mind?" he asked.

"Well, yes, I'd like to think so," I responded while secretly questioning my response. *Did I really?*

I looked at him hopefully, and said, "I'm willing to try anything." It was then that he explained to me that his wife was a distributor for a company that sold a special form of magnet that he thought could be helpful in healing. Flashbacks to my cousin's suggestion of arnica came to mind. But this time, I was humble instead of arrogant. It seemed conventional medicine had no workable solution for me.

This radiologist's wife came into the office that afternoon and brought her magnets. They were two small round discs with a gold foil on one side and a typical magnet-appearing dull brownish-gray surface on the other. She taped one of the discs to each side of my knee and then explained that I should leave the magnets on all the time. If they came off, I should retape them. She explained to me that if I felt increased pain, I could remove them for an hour or two but then put them back on and gradually increase the length of time I wore them.

The instructions were certainly easy enough. I had my doubts, but I respected my colleague and didn't want to be dismissive of something he and his wife seemed to truly believe in. And what the heck did I have to lose? I graciously thanked her and affirmed I would let her know if I noticed any improvement.

I didn't really have any hope that these two small discs would do anything for me, but out of respect for my coworker, I wore them, secretly ruminating about when I should have the arthroscopy and how I should go about canceling my trip. About halfway through my drive home after work that day, I pressed the clutch to switch gears and realized I didn't feel a twinge in my knee. *That's interesting*, I thought.

The following morning, with the magnets still taped to my knee, I awoke, showered, and carefully maneuvered around the house getting ready for work. I managed to avoid any jabs of pain. When I got into the car and carefully pressed down on the clutch, I felt no twinge. I was giddy and felt myself filling with joy. "This is a miracle!" I said out loud this time. I looked down at my knee and laughed out loud. "This is unfreakin' believable!"

When I got home that afternoon, the pain was pretty much gone, so I took the magnets off. But as soon as I did, the twinge came back. Quickly, I reapplied the magnets and breathed a sigh of relief. I obviously wasn't healed–it had only been a couple of days–but these magnets were doing something. Maybe they were affecting my sensory nerves? I wasn't sure.

I wore those magnets for the next two months. I didn't know what they were doing or how they could possibly work, but I wasn't going to take them off again. I wore those magnetic discs to Tanzania and as I trekked up and down Mount Kilimanjaro with its challenging 19,362-foot peak! I also wore those magnets while gorilla trekking in Uganda, up and down steep slopes on uneven terrain, through freshly thrashed brush in the Bwindi Impenetrable Forest. Miraculously, I didn't have any pain, weakness, or instability during the entire trip.

When I got back home two months later, I took the plunge and finally removed the magnets. No pain! I giggled to myself, as if I had gotten away with something. I don't know if it was because I was a physician with a special interest in orthopedic radiology that I was particularly impacted by this healing, but to be healed in this noninvasive manner, for the price of two magnets, with no side effects, no follow-up appointments, and now–thirty years later–no pain or arthritis was life changing. To this day, I am aware that the conventional alternative would likely have been arthroscopy and perhaps a partial meniscectomy (surgical removal of the meniscus, the knee's "shock absorber"), which was the preferred treatment at that time. After surgery, I would more than likely have had scarring in the joint and eventual arthritis that would have plagued me for the rest of my life, possibly leading to a joint replacement. Yes, silly, seemingly inconsequential injuries can cause you instability, pain, and arthritis for the rest of your life if they don't heal properly.

Researching the Health Effects of Magnetic Fields

After this "miracle" cure, I was determined to figure out how these magnets could possibly work. I had been chosen to do a fellowship with a world-renowned orthopedic radiologist in San Diego the following year. His department was a powerhouse filled with clinical fellows and dozens of international research fellows. I had envisioned that, with his backing, I would be able to change the field of orthopedics by proving these static magnetic fields could speed up the healing process! At the time, I was unfamiliar with Robert Becker, whose impactful book, *The Body Electric*, would be released the following year.

Although my mentor didn't doubt that the magnets worked, he carefully expressed his concern that if I began my career with such a project, I could become a pariah of sorts. He cautioned me that medical research moves slowly. Physicians are very reluctant to accept new information that challenges their fund of knowledge. Permutations of existing knowledge are much more easily accepted than the introduction of a totally new concept. To prove that a static magnetic field can alter the healing mechanics of the human body would fly in the face of the biochemical model of human physiology. It would require a *paradigm shift.* Having great respect for his opinion, I acquiesced and instead performed anatomical research. Although I wrote seemingly inconsequential papers on the anatomy of various joints as visualized on MRI examinations, it helped build my credibility.

Fortunately, Robert Becker pushed through and published *The Body Electric*, in which he documents his research uncovering the role magnetic fields have on the body's ability to heal–and in some cases regenerate. Much of his research explored how organisms regenerate limbs, and how healing is affected by the induction of electromagnetic changes in damaged tissue. Although his research didn't directly involve magnets, it was related. Becker's work led to the invention of electromagnetic stimulators, which are widely used in current-day orthopedics to expedite bone fracture healing and bone graft incorporation. Becker's work was revolutionary, but I suspect the true impact of his discovery was lost on most.

Promoting Static Magnetic Fields

Over the years, I have recommended static magnets to many and witnessed a significant number of people heal more rapidly than otherwise expected. When my sister injured her ankle roller skating and endured a fracture that required surgery, she placed the magnets on the outside of her cast. She reported back to me that her surgeon was amazed at how fast the fractures healed. Another friend, a judge, who was scheduled for surgery to repair severe tennis elbow, was able to cancel his surgery because he was cured by applying the magnets to his elbow for a few weeks. One of my department secretaries had radiculopathy (pain radiating down her arm) from degenerative changes in her neck (cervical spine). While waiting for an appointment to schedule an operation with a neurosurgeon, she was willing to wear a magnetic necklace for a few weeks and during that time, slowly became symptom free. She no longer needed surgery.

These impressive results are not achieved in every person or for every kind of injury or source of pain. However, over the past thirty years, I have witnessed so many successes in those who are skeptical and hesitant about even trying them, that I know the benefits of these products are not simply due to a placebo effect.

The magnetic products I have used are produced by a Japanese company called Nikken. To be clear, the Nikken products aren't simple bar magnets you can buy in a hobby shop or toy store. A bar magnet produces a permanent, static magnetic field, meaning, the magnetic field doesn't change over time and lines of force are closed. As far as I know, a simple bar magnet will not heal an injury. Nikken magnetic products, on the other hand, are much more complex, with varying polarity along the face of the disc surface resulting in a static, three-dimensional complex magnetic field. This is an energy-based technology and therefore does not constitute "medicine" in the traditional sense and does not provide a "medical treatment." Perhaps these magnets can be considered a form of energy manipulation.

How Can Localized Magnetic Fields Heal?

Despite my long history of using and promoting magnets to speed up healing, I still do not fully understand how these products work. What I do know is life has evolved within the envelope of Earth's magnetic field, which is extremely weak, between 250 and 650 milliGauss (25-65 µT). Our bodies are designed to respond to subtle magnetic fields. When a magnet is applied to the skin, the magnetic field penetrates through the body's tissues creating a localized magnetic microenvironment. The magnetic discs are a composite of numerous mini magnets arranged in a specific and complex design, which produces a three-dimensional magnetic field with regions of varying polarity and intensity, which penetrate deeply into the body. Given that the body is filled with charged ions and water molecules, which themselves function as mini dipole magnets, it seems likely that when flowing blood passes through the changing magnetic field, these charged particles may flip back and forth, subtly dilating capillaries and increasing blood flow. It would make sense that more blood in and more blood out enhances body tissues' ability to heal.

Magnetotherapy reduces inflammation. Although an immune response is designed to heal, the aggregation of cells caused by an immune response can cause swelling and increasing pain, and in some instances delay timely repair. By allowing more cells and inflammatory proteins to access an injured area and, at the same time, enable more clearance of debris, these products seem to create an environment that helps the body heal itself. I have witnessed fractures heal in less than half the expected time. Tendinitis and bursitis can resolve relatively quickly without the use of anti-inflammatories. Even hematomas can dissipate much more rapidly when placed in a static magnetic field.

There are limitations to what static magnetic fields can repair. If a structure is strained or sprained, it may benefit from a brace, but the repair can be sped up with magnetic fields. If a tendon completely tears and crinkles up–a process known as *retraction*–a static magnetic field will not help the tendon repair. Retraction commonly occurs with rotator cuff tears, biceps tendon tears, and tendon tears in the foot and ankle. In general, the proper healing of something that is torn or broken requires the contiguity of the damaged structure's pieces, which is no longer the case when a tendon retracts. Fractures may need to be stabilized and immobilized by a cast or splint in order for the fragments to heal and knit together. Depending on the body part and the number and location of bone fragments, stabilization may require fixation with metal plates and screws. Without immobilization, the movement of bone fragments during the recovery process will delay healing or even prevent the proper fusion of fracture fragments. Magnets will speed up the healing process, but they do not provide immobilization. They can be a helpful adjunct after immobilization and/or surgical fixation. The same holds true for ligament injuries. If a joint is unstable because ligaments are torn or stretched, a magnet cannot stabilize a joint. But, if the joint is stabilized with a brace, or if surgery is required, a magnetic product may help speed up repair.

Over the years, my medicine cabinet has gradually shifted from one filled with anti-inflammatories and pain relievers to one filled with various magnetic products. I use magnets for sprains, strains, and bruises on myself and on my children. I can't even remember the last time I ingested an anti-inflammatory. It seems like another lifetime when anti-inflammatories were my "go to" medication for pain. For the most part, magnetic products are worry free, except in people with a history of arrythmia or in those with an implantable device. Cardiac pacemakers, defibrillators, spinal cord stimulators, or other devices could be reset by a magnetic field, so those with implants should be careful using these products and keep them far away from any of these devices.

Although many vendors sell magnetic products, this is a technology and there are differences among magnet design and field strength. Stronger magnets don't translate into a greater healing effect and may not be bioactive. Strong magnetic fields can also be detrimental in some instances, as previously described.

Healing with Electric Fields

Pulsed Electromagnetic Fields

Observations that magnetic field exposure can produce beneficial health effects led to the therapeutic technique called *pulsed electromagnetic therapy* (PEMT). As the name implies, PEMT is a type of therapy that involves the generation of pulsed, repetitive electromagnetic fields that are directed into the body.

The technique and parameters used to create pulsed magnetic fields vary among vendors and practitioners offering this technology. The frequency and amplitude of the magnetic field, in addition to the spacing between pulses, are variables that can affect the therapeutic value of the product. Numerous researchers have explored varying parameters to determine the impact of each on the patient experience. For example, in one study, a magnetic field with an amplitude of 2 tesla, a frequency range of 1-50 Hz, and a pulse width of 270 microseconds was associated with a decrease in back pain.[274] In a prospective, placebo-controlled, double-blind study, PEMT with magnetic field frequencies between 4 and 12 Hz and an amplitude of 105 mT was shown to significantly reduce knee pain in patients with osteoarthritis.[275] PEMT has been used as an anti-inflammatory and, like static magnetic fields, can dilate microcirculation and increase oxygen and nutrient delivery to the body's tissues, including the brain.[276] PEMT does not improve lymphatic flow.[277]

This technology was approved in 1979 by the FDA as a method to control pain in many different clinical settings. Despite its approval forty-five years ago, it remains a controversial therapy. A popular vendor for PEMT is BEMER. For some conditions, particularly in patients with poor circulation, these products can be extremely helpful.

Electrostimulation (E-Stim)

Another technology that utilizes electromagnetic fields to promote healing is *electrostimulation*. This technology creates localized EMFs and has been used in conventional medicine for decades, ever since Dr. Becker demonstrated their beneficial effect. This device, called an E-stim, consists of a small battery pack with attached electrodes that are placed within nearby soft tissue structures. E-stims have been used to accelerate the healing of fractures that may otherwise be slow to heal. They have also been used to accelerate the surgical fusion of joints, a procedure called arthrodesis, typically reserved for severely damaged, end-stage joints that produce chronic pain and/or instability.

E-stims are often placed into the spinal canal, where they provide stimulation to the spinal cord to help relieve chronic pain syndromes. An advancement of this technology, termed functional E-stim (FES), targets specific nerves to activate muscular contraction and allow for movement that otherwise would not be possible.[278] The advances in this field are truly amazing.

A Hair of the Dog

Electrostimulation is also used in the brain. Transcranial placement of leads, i.e., leads placed through the skull, are directed deep into the brain and can sometimes be helpful for those with neurodegenerative disease. Conditions such as Alzheimer's disease are sometimes worsened by, or brought on by, strong, ambient EMFs. How interesting then that a potential treatment for this disease is the placement of leads into the brain so electromagnetic pulses can be delivered right to the brain and *reduce* the symptoms of this debilitating disease.[279] Long-term use of weak electromagnetic fields has also been useful in treating patients with multiple sclerosis and optic neuritis.[280]

Another therapy, called *transcranial magnetic stimulation* (TMS), is a noninvasive technique during which magnetic fields are applied to specific regions of the scalp to induce electric currents in different parts of the brain. TMS can be used to treat many conditions, including Alzheimer's, ALS, tinnitus,[281] depression,[282] and many others. This therapy can be used to treat seizures in patients with epilepsy,[283] but in some individuals, depending on the frequency of the magnetic field delivered to the brain, TMS can *cause* seizures.[284] As it seems to be a general rule with electromagnetic radiation, the thin lines between a beneficial effect and a harmful effect are related to frequency, intensity, and individual susceptibility.

Frequency Specific Microcurrent (FSM)

A complementary technique called *frequency-specific microcurrent* delivers microcurrents (one millionth of an Ampere) of electrostimulation to parts of the body needing repair. The specific frequency chosen depends on the specific condition being treated. This procedure is FDA approved and has been shown to be effective at relieving chronic pain and promoting healing. The full breadth of this technique's utility is not yet known but is actively being researched.[285] Although the currents are tiny and cannot normally be felt, unless the intensity is turned up high, they do travel through the body's tissues, and those individuals with pacemakers and other implantable devices may not be able to undergo this treatment.

Ablation

The word *ablation* means to remove or destroy a body part or tissue. Ablation is most commonly performed surgically.

However, ablations can also be performed by exposing the body to EMR. Ionizing radiation can be used to ablate tumors and is also a common method for destroying the thyroid gland in patients with hyperthyroidism. Ablation is also commonly performed with nonionizing radiation, by directing RF radiation to unwanted tissues. One of the more common applications of RF ablation has become treatment of arrhythmia.

The heart's conduction system enables the synchronous contraction of the heart. When the conduction system doesn't function properly, the heart muscle can flutter or *fibrillate* (a very fast and ineffective twitch), a condition called an *arrhythmia*. There are several different types of arrhythmias, but the most common is atrial fibrillation, or Afib. This condition is dangerous because a prolonged arrhythmia can lead to blood stagnation and clotting. If blood clots are ejected from the heart and into the arteries, they can cause strokes and block blood vessels supplying the organs and limbs, leading to tissue damage and possibly death. Arrhythmias can sometimes be controlled with medications but may require an invasive procedure, such as the placement of a pacemaker.

RF ablation is a relatively new procedure used to treat people with arrhythmias. This procedure was initially described in the mid-1980s and has become increasingly common.[286] This relatively less invasive procedure spares the patient having to undergo a cardiac pacemaker implantation. During cardiac ablation, a catheter is placed into the heart. Then, high-intensity RF radiation with a frequency between 500 kHz and 1 MHz and a power density over 40 W is aimed at the atrioventricular node. This is the part of the heart's conduction system that is often faulty and responsible for the abnormal rhythm in patients with Afib. The RF radiation heats and destroys the tissue, often resolving the arrhythmia.

RF ablation has also become an alternative to surgical resection of tumors, both benign, like uterine fibroids, and malignant, including kidney, liver, brain, and other cancers.[287] By precisely localizing and controlling the intensity of the radiation, tumors can be melted, leaving the normal organ structure around the tumor intact. The technology has dramatically changed the practice of surgery, and in particular, surgical oncology. Patients heal faster, and there is much less overall stress to the body as compared with a similar "open" procedure.

Diagnostic Imaging

Although most of the machines in the typical radiology department use ionizing radiation to generate images, by using non-ionizing radiation, magnetic resonance Imaging (MRI) can generate exquisite images of the body's interior. The physics behind MRI is complex, but can be divided up into two components. The first is the immersion of a body into a strong magnetic field. The other is the reoccurring, periodic exposure of the body to pulsations of RF radiation.

Magnetic field strengths created by an MRI machine vary depending on the machine. When the body is placed within the magnetic field, a small percentage of protons in the body align with the magnetic field. Higher-field strength magnets increase this percentage and produce sharper images. Most commonly, MRIs generate a magnetic field strength of 1.5 tesla. Some newer machines produce a field strength of 3 tesla. Portable extremity units often produce images using the lowest field strength, which may be as weak as 0.2 tesla.

The second part of the MRI experience involves exposing the body to pulsed RF radiation. An RF radiation pulse will travel through the body and knock protons that are aligned with the magnetic field off axis. This is a transient phenomenon, as the protons will then revert back and realign with the magnetic field. When they do, the protons emit a frequency of EMR that receiver coils placed on the body can detect. Depending on the frequency of the RF pulses, which typically range from 1 to 300 MHz, and other parameters, the signal emitted by the body's tissues will vary. The computer's software can listen for and localize the source and quality of each proton's emission, interpret variation, and create images. A typical MRI exam will consist of several different protocols or sequences to create different-appearing images. This helps the radiologist separate out tissues with different characteristics and define the anatomy or structure of the part being studied. It is a very sophisticated technology, to say the least, and one that has revolutionized the entire field of medicine.

The biggest obstacle to tolerating an MRI is claustrophobia. On occasion, a patient will need to take anxiety-reducing medication to overcome this phobia. But for most people, MRI examinations are easy to tolerate and are safe. I've heard some say they feel a subtle sense of confusion after having an MRI examination, but the deficit doesn't last long. Patients are screened to ensure they don't have any fragments of metal in their body that could shift when in the presence of a strong magnetic field and cause damage, such as in the eye. Cochlear implants and some other implantable devices can also be contraindications for MRI.

Most MRI exams are performed without contrast agent. However, depending on what the study is being done for, a contrast agent may be needed for better definition of normal from abnormal tissue. Good kidney function is required before MRI contrast agent can be injected into the body The rare earth metal gadolinium is used as a contrast agent for MRI examinations. Although companies have performed stability tests on gadolinium compounds, I have heard from clinicians who perform chelation studies that free gadolinium is often picked up on these tests after a patient has had an MRI with contrast. If you have had an MRI with contrast and want to rid your body of the gadolinium that might have become dissociated, chelation is a viable option.

Healthcare Practitioner Training

Patients who present with physical complaints caused by EMF exposure are currently treated symptomatically, but the root cause of their disease–EMF exposure–is generally not part of the thought process and therefore not in the differential diagnosis. To remedy this, many experts in this field focus on educating and training healthcare practitioners to understand and recognize the health effects of EMFs. They offer medical conferences to train professionals about the health effects of EMFs and RF radiation. Free lectures for physicians from a 2021 EMF medical conference are available on Vimeo at https://vimeo.com/showcase/10624511. The first medical text-book–*Electromagnetic Fields of Wireless Communications: Biological and Health Effects,* edited by DJ Panagopoulos, PhD and published in 2022 by CRC Press–is now available. In addition, *The Alliance of Nurses for Healthy Environments Textbook: Environmental Health in Nursing, 2nd Edition,* includes the chapter "A New Form of Environmental Pollution: Wireless and Non-Ionizing Electromagnetic Fields" (pages 136-60). This book can be downloaded for free at https://bit.ly/440pG6B. It will take time to educate the medical community, but having reputable resources will certainly help.

CHAPTER 14

PROTECTING OUR CHILDREN

When we look to our children, we see the future. Regardless of whether you raised a family, the younger generation will eventually need to run our world. We can dream for evolution and for a more enlightened and just civilization. Maybe they will do a better job? In order to advance our society, it is truly important we educate our children and ensure they are healthy, well-adjusted, grounded, and connected to one another and to us.

Like adults, children are constantly being exposed to non-ionizing radiation. These include all of the categories described in this book: ELF from power lines and household wiring, VLF and LF from dirty electricity, and RF/MW radiation from wireless communication, and now MMW from 5G. But children are different from adults in that they need to continue growing and developing, in some cases through their early twenties. It is important to understand how sources of EMF exposure can affect their well-being. Let's start with pregnancy, as the effects of ELF and RF/MW radiation appear to start when the baby is still in the mother's womb.

ELF-EMF and Pregnancy

Children may experience negative health effects from EMFs even before they are born. Because ELF-EMF penetrates deep into our bodies, the proximity to sources of ELF, such as the location of power lines, should be considered when choosing a house or apartment. Although you may not have heard about the danger of high-voltage power lines in a long time, the problem has not gone away. In fact, research is still ongoing. If you are planning on having a family, you may want to choose a home that is far away from high-voltage power lines. The distance considered safe is variable and should ultimately be determined by performing an indoor survey of electric and magnetic field strengths. In a study from 2017, researchers determined that magnetic fields over a threshold of 2.5 milligauss (mG; 250 nT) have a 48 percent increased risk of miscarriage than with lower exposures.[288] Research has also shown there is an increased risk for premature birth when the expecting mother lives closer than 600 m (1,968.5 ft) to a high-voltage (230 and 400 kV) power line.[289] Although some have questioned whether there is an increased rate of birth defects in children born in close proximity to high-tension power lines, results from studies performed during the past ten years are inconclusive.[290, 291]

In addition to potential complications during pregnancy, increased prenatal exposure to ELF (magnetic fields in particular) may also be causing an increase in the prevalence of chronic childhood diseases. When mothers wore detectors while carrying out daily routines during pregnancy, researchers found asthma was more common in children (up to thirteen years of age) of mothers who were exposed to higher magnetic field levels.[292] In fact, the authors of this study observed a linear relationship in which every 1 mG (100 nT) of increased magnetic field exposure during pregnancy was associated with a 15 percent increased rate of asthma! If the mother had a field exposure of > 2 mG (> 200 nT), their children had more than a 3.5-fold increased risk for asthma when compared with the children of mothers with lower MF exposure.[293]

Childhood obesity is another chronic health condition research has linked to increased magnetic field exposure during pregnancy. In another study, pregnant women wore ELF-EMF monitors while doing their normal routines in the first trimester. Children who were born to women exposed to high ELF-EMF between 40 and 800 Hz (including 50/60 Hz electricity) had a 69 percent increased risk for developing obesity as compared with the children born to mothers who had low ELF-EMF exposures.[294] In a later study, the same research team documented an association between fetal magnetic field exposure and increased risk for attention-deficit/hyperactivity disorder (ADHD), but this paper was later retracted after the statistics were called into question by one of the journal's readers. The authors were asked to rework their data, and when they did, they found that although there was still an association of MF exposure and ADHD in some of their subjects, the associations were inconsistent and nonlinear.[295]

Getting into the nitty-gritty of research protocols and statistics will often reveal technical and procedural flaws that challenge the validity of the study results. In some cases, the critique is justified. In other instances, industry influence on journal editors can cause good papers to be retracted, keeping the public in the dark. That is how it goes in science. Researchers need to prove or disprove the findings of others, with the goal being that truth will ultimately be revealed. Based on what I've read to date, I recommend that pregnant women limit their magnetic field exposure. Two mG (200 nT) is not an uncommonly high magnetic field level, particularly if your home is next to a residential power line. The only way to know the intensity of magnetic fields in the home is to survey it yourself or to hire a building biologist. If you want to do it yourself, consider these sites for guidance: EMF analysis.com and safelivingtechnologies.com.

In addition to carefully choosing the location of your home, there are several products in the marketplace that pregnant women should avoid, as they can unknowingly expose themselves and their future offspring to strong magnetic fields. Think electric heating pads and electric blankets. Chairs and sofas that plug in may also generate magnetic fields. Test the field intensity of these appliances before using them. In general, avoid placing anything that plugs into an AC outlet on a pregnant belly or on the back of a pregnant woman for prolonged periods of time.

RF and Pregnancy

ELF-EMF is not the only form of radiation pregnant women should avoid. Modulated MW and MMW radiation used for data communication, including 5G networks, may penetrate the body's tissues. Even though the fetus is surrounded by amniotic fluid, it may be susceptible to EMR penetration by the formation of Brillouin precursors and wave dispersion. I read an interesting scientific paper and review in which the authors concluded smartphone radiation was associated with an increased risk of miscarriage, in addition to changes in heart rate variability and hormonal changes in the mother. This article, too, was quickly retracted by the journal. In this case, the journal editors specified they identified several critical problems impacting the article's reliability, including a "narrow view of the published literature related to the topic and an incorrect conclusion." The reason for this retraction seemed more dubious.[296] A different paper studying the effects of cellular phone use on fetal heart rate showed that HRV in the fetus decreases with mobile phone use by the mother.[297]

Increased fetal stress and other nebulous effects from radiofrequency and microwave radiation may have an impact on the child that could surface years after birth. Researchers in Yale's Department of Obstetrics, Gynecology and Reproductive Sciences found that pregnant mice exposed to RF radiation gave birth to mice that grew up to be hyperactive with impaired memory, a condition we humans would label ADHD behavior.[298] A very large prospective study that looked at over eighty-three thousand mother-child pairs found that high prenatal cell phone use by the mother was linked to hyperactivity and inattention problems in their offspring, similar to the effect of magnetic field exposure during pregnancy. If a mother did not use a cell phone during pregnancy, her children had a low risk of future hyperactivity/inattention or behavioral problems as reported by their mothers through the study's child behavior checklist.[299]

A study of three birth cohorts looked at the effect maternal cell phone use during pregnancy had on the child's cognition at five years of age. The authors found that the children of women who frequently used cell phones during pregnancy had lower mean cognition scores than those children of mothers who did not. The authors were cautious to clarify that they couldn't definitively determine whether this effect was truly caused by prenatal cell phone use or other factors.[300] Further research should be performed to investigate this possibility.

We live in a society currently immersed in microwave radiation everywhere we go, and expectant mothers can't be removed from daily exposure. But if pregnant women limit cell phone use, turn off the antennas not in use, and keep active cell phones away from their bellies, their babies' exposure will significantly drop. In addition, pregnant moms should not place a laptop, tablet, or active cell phone on their belly and should avoid wearing smartwatches, Bluetooth devices, and any other device that sends and receives information wirelessly. Creating an RF-free sleep sanctuary is critically important too. See the Babysafeproject.org.

Newborns

Once a baby is born, it is no longer physically surrounded and protected by its mother. Newborns are sensitive to EMFs. A study by Bellieni et al. showed that heart rate variability in newborns placed in incubators dropped when the incubator motor clicked on, indicating a stress response.[301] Although it may seem logical that an infant placed in an incubator would be under stress, scientists observed that the HRV consistently returned to normal, indicating the stress response ended each time the incubator's motor clicked off. In addition, the researchers determined that the stress effect was not related to the clicking or whirring *sound* of the motor, narrowing the stress-inducing agent to the ELF-EMF produced by the motor itself.[302] If and when companies understand the impact EMFs have on an infant's well-being, hopefully they will create incubators in which the internal environment is electromagnetically clean. Monitoring will then take place without unnecessarily increasing the infant's stress level. Hopefully, this advancement in technology will soon be developed and implemented.

Baby at Home!

Bringing a newborn home can be a scary proposition, particularly if the parents are young and it's their first child. What do you do when the baby cries? What if you don't hear the cry? Should the baby sleep in the same room as the parent(s) until it is stronger? These are just a few of the many questions a new parent may have.

Most every baby registry includes a monitor in the list. When the newborn first gets home, many parents immediately plop a baby monitor down next to their sleeping infant to ensure that they'll know if the baby is asleep or awake, cooing or crying. Aside from providing a false sense of security, baby monitors emit ELF-EMF in addition to RF/MW radiation within a frequency range of 400 MHz to 2.45 GHz, depending on the vendor. Several studies have assessed infants' exposure levels from monitors, but exposure naturally depends on how far the monitor is placed from the baby. Like all RF radiation sources, the closer the emitter, the more intense the radiation. In addition, if the parental unit is further away from the baby's unit or placed in such a position that the communication between the two units is compromised, the infant unit will increase its transmission power density in an attempt to maintain connection, increasing the baby's exposure. Manuals usually specify that a monitor should be placed at least one meter from the infant's bed to reduce exposure to ELF-EMF fields from the device. However, many parents don't pay attention to this detail. Furthermore, ELF-EMF intensity will be *reduced* with a one-meter distance between monitor and baby, but this is not enough of a distance to *minimize* RF/MW radiation exposure.

If you use a baby monitor, keep the monitor as far away as possible from the baby and consider separating the monitor from the baby even more with shielding material. Although it might seem like a good idea to drape an EMF-shielding fabric over the crib, if it should come down and get into the crib, it could put the infant at risk for suffocation. So, please take special precautions not to have any loose fabrics around the crib.[303] A simple solution could be to place a large piece of cardboard covered with aluminum foil between the baby monitor and the crib. In this way, RF from the monitor will bounce off the foil, limiting the baby's exposure, while at the same time allowing sounds from the baby to reach the monitor.

The more physical structures placed in the pathway between the parental unit and the baby monitor, the more the monitor will increase its signal strength to achieve communication. Consider this when trying to decrease your baby's exposure as much as possible. Another option is to only use the monitor when needed and, instead, have the baby's crib in close proximity to the mother. The sensory connection to mother is extremely important to give the baby comfort and a sense of security. For more information, check out the fact sheet available from The Building Biology Institute at: https:// buildingbiologyinstitute.org/free-fact-sheets/baby-monitors.

Protective measures should be taken to minimize a baby's exposure to other RF/MW radiation in the home.[304] Place the Wi-Fi router as far away from the baby's room as possible. Avoid bringing cell phones into your infant's room, and try to limit cell phone use while cradling the baby. In addition, consider placing cell phones on airplane mode when taking pictures and videos. Keep wearables in airplane mode too.

Children

When my kids were little, there was no Wi-Fi in our home and no tablets, laptops, or smartphones. As technology has evolved, methods of entertainment for children have dramatically changed. Now, small kids are often handed a screened device equipped with Wi-Fi. As we've discussed, these devices, held by little hands close to the body, emit RF/MW radiation and blue wavelengths of light. Sometimes, kids are plugged into these devices at home, while in the car, at school, and at restaurants. Regardless of the temptation, it would be best to keep RF/MW-emitting devices away from your children as much as possible.

How old were your children when they got their first cell phone? How about their first tablet? Research has shown that cell phone use by children can have health repercussions. A study of over fifty-two thousand children found that those with cell phone exposure in the womb and as children were 30 percent more likely to suffer from migraines and other forms of headaches than those children who were not exposed.[305] Furthermore, seven-year-olds in this group had the highest percentage of behavioral problems as determined by Age-7 Questionnaires.[306] Children who spend more time on their phones, even with the 3G frequencies (900 MHz and 1800 MHz), have disturbed sleep. Symptoms may include shorter sleep duration with night awakenings, abnormal movements, and sleepwalking and talking during sleep.[307]

We held out for as long as we could, but when our kids turned thirteen, they received their first smartphones. Although I thought it was too early for each of them to be handling such an expensive "toy," all of their friends had mobile phones and I realized the value of "keeping up with the Joneses" for a young teenager. In order to try and limit their daily exposure, despite the fact that we called them "their phone," they were only allowed to use them during a few designated hours of the day. The phones were not allowed at the kitchen table during mealtime. At the end of the day, the phones had to be put in airplane mode and then placed into a hallway basket a half hour before bedtime. Our children enjoyed their phones and surprisingly never complained about not having enough phone time. Importantly, we all enjoyed our time interacting with one another and together as a family, particularly at mealtime. I believe this is in part why they both developed into engaging, mindful young adults rather quickly.

What Really Happened to Our Children During the Pandemic?

During COVID, our kids were suddenly at home 24/7 and planted in front of screens. Like many of you, I witnessed my two teenagers sitting in front of their respective laptops during remote school hours dozing, eating, turning off the camera, and muting the device, in order to have conversations with each other and classmates on their cell phones. It was distressing, as I knew these were crucial years for growth and development.

When the pandemic hit, the house policy had been to turn off the Wi-Fi router at night, but during the day, it was left on. I had not fully hardwired the house with data ports for internet access. But after a few weeks of online schooling, I began to hear complaints from my kids that their heads hurt. That spurred me into action. First, I purchased blue light-reflecting glasses for all of us to reduce eye exposure to the excessive blue light wavelengths radiating from the LED screens during the day. Then, I purchased ethernet cables and adapters for each laptop so we could directly plug each device into the back of the router. I had our internet provider dismantle the Wi-Fi capability of the router and made sure the "hot spot" feature was also disabled. From then on, the Wi-Fi router was *almost* always turned off. We had ethernet cables strewn all over the floor, which wasn't the tidiest way to handle the situation, but the headaches stopped. It took me some time to get an electrician over to the house, but after people were going back out and working, I had him install data drops on each floor of the house, into which I connected ethernet cables from the router. After that, cords were better concealed. Bluetooth and Wi-Fi settings on the laptops have been turned off ever since, unless we are traveling.

It was costly to hire an electrician to install the data drops, but the effort paid off. I never heard about headaches again. My kids didn't suffer any anxiety or depression. They were playful and seemed happy. In fact, they enjoyed the time at home during the pandemic and had very few complaints of any kind. They missed the physical presence of their friends, but from the outside, their mood and activity levels seemed normal. They communicated clearly, and their memory seemed normal, which was not the case for everyone. An adolescent's memory can also be affected by devices emitting RF radiation. In a cohort study, RF radiation exposure from mobile phones was associated with an adverse effect on cognition, particularly involving figural memory.[308] I was very impressed at how removing wireless technology from the house favorably impacted my children. Although they were frustrated with not having access to Wi-Fi, they seemed to understand on some level they were benefitting in some way from hardwiring their devices.

Midway through the pandemic, my daughter was invited on a two-week-long camp retreat with her campmates during the school year. It was an opportunity for them to be in each other's presence and socialize. This helped them get through the monotony of those days. Upon her return home, she told us an interesting story. While at camp, she found herself at a table with three other girls who were doing their schoolwork. They each had a laptop connected to the camp's Wi-Fi. In addition, each girl had a cell phone, no doubt with data, Bluetooth, and Wi-Fi antennas enabled. While working at the table, my daughter said she started to notice the onset of a dull headache. On her own, she picked herself up and moved to another table, far from the group, where her headache quickly resolved. When she finished her story, I tried to conceal my smile. I found it encouraging that despite her impatience with my "out-of-the-box" rules regarding the Wi-Fi router, she understood their importance.

Teenage Angst and Depression

Like many, my parenting skills were ad hoc, as I occasionally used intuition to make decisions and oftentimes took lessons from others. I feel fortunate my kids have gone through most of their adolescence without too much trauma. But I've witnessed anxiety and depression in many of their friends. Their experience is not unique. News anchors continually report children are depressed and committing suicide at an alarming rate. Even the surgeon general has recently raised an advisory regarding children's mental health. Questions are being asked, such as, "Isn't there anything we can do to prevent this?" Fingers are pointing at social media. *It's because of bullying. No, it's because of COVID and social isolation. It's because of too much screen time. It's because of the vaccines!* Perhaps these influences have had some impact, but I am most concerned today's kids are suffering the health effects from changes to their electromagnetic environment. And I'm not alone in that suspicion.

As a teenager, I spent a lot of time on the phone. But now, talking on the phone means being exposed to high-intensity RF/MW radiation for prolonged periods of time. I've told my children on several occasions that spending too much time on the cell phone can cause them to feel irritable and anxious, and even depressed. I believe it made an impact, as on more than one occasion my daughter told me her friends were depressed. I initially thought it was the usual teenage angst, but when she added that they were on medication, I became concerned. One of our conversations went something like this:

"How much time do they spend on their phones?"

"They are all constantly on their phones. I tried to tell them to turn them off before they go to sleep, but they won't. My one friend even sleeps with her phone under her pillow!"

My daughter replied with sadness, as she loved her friend group, but they didn't understand or even want to know what she was trying to tell them.

One night I had the opportunity to observe my daughter's friends in action. I had agreed to let her host a Halloween costume party. It was unseasonably warm, and I set up a pre-party dinner in the backyard for five of her closest friends who were over early to help set up. They were excited, talking about their costumes, makeup ideas, and things they still needed. I listened with silent pleasure, reminded of the crazy Halloween parties I had gone to when younger. At one point, one of the girls took out a small container and popped a pill into her mouth. "It's my Zoloft," she said. "Lexapro does nothing for me anymore." Another girl made a joke about it and took two of her own pills. They then shared details of their journeys on antianxiety meds. The conversation was bizarre. I found it hard to grasp that these "kids" thought their need for medication was normal and even fodder for jokes.

Feelings of sadness and depression certainly occur in teenagers, particularly in girls. This is nothing new. In a meta-analysis of epidemiological studies performed from the 1960s through the mid-1990s, approximately 5.9 percent of thirteen-to-eighteen-year-old girls and 4.6 percent of the same-aged boys experienced depression during those decades.[309] That translates to one child with depression in a classroom of twenty students. During that thirty-year time span, there had been no statistical change in the prevalence of adolescent depression.

But in the mid-1990s, professionals began to raise the alarm that an increasing number of children and adolescents were suffering from depression. The general consensus at that time, though, was that the perceived increase in the numbers of affected children was due to a greater awareness of childhood depression and not a true rise in numbers of those with the disorder. But during that time, the use of wireless technology and therefore children's exposure to RF/MW radiation began to escalate. In 1997, Wi-Fi became available to the public and in 2007, the first smartphones came to market. During that time, the prevalence of adolescent depression (ages twelve to twenty years) began to significantly rise. In 2005, 8.7 percent of children were depressed, and in 2014 this statistic increased to 11.3 percent, i.e., one in nine![310]

When statistics from the Substance Abuse and Mental Health Services Administration (SAMHSA) came out for 2020, their data indicated that 21.9 percent of adolescents had suffered a major depressive episode. One in four adolescent girls (25 percent) had at least one episode of major depression in 2020, as compared with one out of six boys.[311] Some attribute this dramatic number to the COVID-19 pandemic, and there was likely an impact from COVID for sure, but the percentages of children with depression were already going up before COVID (Figure 14.1). Teenage depression has increased in concert with the usage of mobile phones and wireless technology.

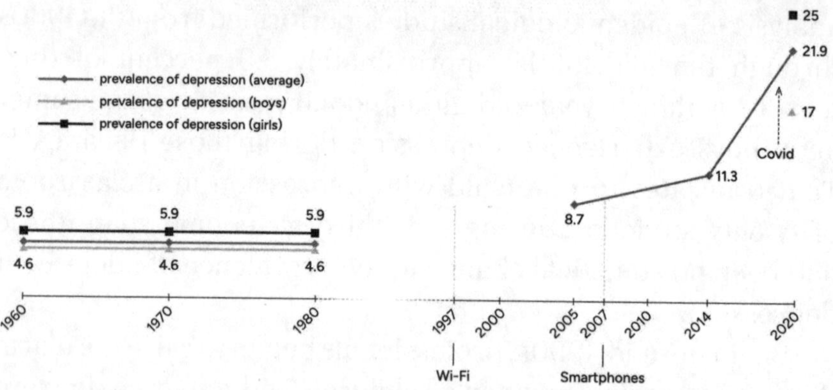

Figure 14.1. Prevalence of adolescent depression from 1960 to 2020

Certainly, there are a myriad of reasons more adolescents are becoming depressed. The threats imposed by climate change, the contamination of the food supply, dyes, processed foods, family financial woes, entertainment sources, and social media have all have a role. But some epidemiologists have associated cell phone and laptop use with adolescent depression. Content has been blamed for causing anxiety and depression in children. Bullying on social media, violence in video games, observations of celebrities who are unkind, and watching other kids harm themselves may be nudging an uptick in clinical depression among our children. A literature review studying the adverse effects of screen time on children and adolescents determined that digital media can have adverse physical, psychological, social, and neurologic consequences. This review affirmed a link between overall screen time watching violent, fast-paced content, with depressive symptoms and suicidal ideations, particularly if devices were used at night.[312] An important book recently published on this topic, called *The Anxious Generation* by Jonathan Haidt, cataloged the multitude of ways in which addiction to screens and excessive screen time has damaged an entire generation of children. It is an excellent book which I highly recommend.

Although content may be in part to blame, research has also shown that screen time alone may be a risk factor for both anxiety and depression in adolescents.[313] Some suspect content is irrelevant, but it has been difficult to separate out the effects of the modulated RF/MW radiation itself from its content. It would be interesting to compare the risk for anxiety and depression in children who use hardwired devices for social media with their peers who rely on wireless technology.

Gaming controllers, wireless speakers, Bluetooth earbuds, virtual reality headsets, and many other devices emit significant levels of EMF, but their intensities are lower than the radiation emitted by Wi-Fi routers and cell phones. Cell phones potentially expose children to the highest intensity of RF/MW radiation at home. But thankfully, texting and video chats have become much more common, exposing the brain to significantly less intense radiation as compared to a mobile phone placed against the ear during a voice call. Other devices, however, are being designed, such as Bluetooth headphones and virtual reality headsets, which expose the brain to increased intensities of RF/MW radiation.

Children have thinner skulls than adults, which are more easily penetrated by EMFs. Figure 14.1 was taken from a paper written by Dr. Om Gandhi. There are many other similar images floating around the internet, which all document the same idea. Cell phone radiation affects brain activity, and this effect is more dramatic and can even cross midline in children and affect the other half of the brain. Any RF/MW-emitting device applied to the head will likely have a similar effect.

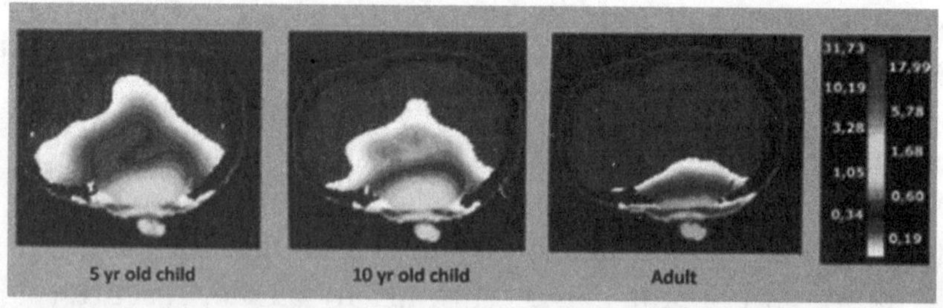

Figure 14.2. Absorption of MW radiation by the brain according to age Courtesy of Om Gandhi, et al. *Electromagnetic Biology and Medicine*, 31(1):34-51, 2012. Taylor & Francis Ltd, http://www .tandfonline.com

A modeling study by Fernández et al. showed that the absorption of radiation from virtual reality headsets is also much more substantial in a child's brain and eye than in an adult.[314] The level of intensity may not be high enough to cause cataracts, but more subtle effects on mood and behavior are certainly a concern if these devices are used for prolonged periods of time. As virtual reality headsets become more popular, they will become a prevalent source of significant RF/MW radiation exposure to the brains of the younger generation.

What Does the American Academy of Pediatrics Think?

For those who think this information is alarmist and out of sync with mainstream medicine, consider these recommendations for children and their families first published in 2018 by the American Academy of Pediatrics on their healthychildren .org website:

- Use text messaging when possible, and use cell phones in speaker mode or with the use of hands-free kits.
- When talking on the cell phone, try holding it an inch or more away from your head.
- Make only short or essential calls on cell phones.
- Avoid carrying your phone against the body like in a pocket, sock, or bra. Cell phone manufacturers can't guarantee that the amount of radiation you're absorbing will be at a safe level.
- Do not talk on the phone or text while driving. This increases the risk of automobile crashes.
- Exercise caution when using a phone or texting while walking or performing other activities. "Distracted walking" injuries are also on the rise.
- If you plan to watch a movie on your device, download it first, then switch to airplane mode while you watch in order to avoid unnecessary radiation exposure.
- Keep an eye on your signal strength (i.e., how many bars you have). The weaker your cell signal, the harder your phone has to work and the more radiation it gives off. It's better to wait until you have a stronger signal before using your device.
- Avoid making calls in cars, elevators, trains, and buses. The cell phone works harder to get a signal through metal, so the power level increases.
- Remember that cell phones are not toys or teething items.[315]

Many teachers and educators are creating school policy to limit cell phone use in schools. Teachers' unions are taking steps to reduce emissions in schools to lessen exposure as much as possible. At the same time, there are school administrators who have contracted with wireless companies to install antennas on school property, significantly increasing the exposure of everyone within the building.

What Can You as a Parent Do Now?

At the present time, there are a number of pioneers doing important research and trying to get the word out to the public. Our legislators and policymakers are largely unaware of these health issues. But this is slowly changing.

New Hampshire became the first government in the country to investigate and came out with a groundbreaking report documenting the conflicts of interest with the industry and government agencies. My colleague and friend, Kent Chamberlin, PhD, a professor emeritus and former chairman of the Department of Electrical Engineering at the University of New Hampshire and former president of the Environmental Health Trust, sat on the commission. Their final report (HB 522, Chapter 260, Laws of 2019, RSA 12-K:12-14) makes fifteen recommendations to transition away from wireless technology to hardwired solutions to and through the premises.[316]

In 2022 Louisiana successfully passed HB548 requiring the state Department of Education to develop and distribute health and safety guidelines relative to best practices for the use of digital devices in public schools.[317]

I also received an email notification that a Maine state representative has become interested in the health effects of wireless technology and has written a bill to test the long-term effects of 5G radiation on children. Why did she take interest in this topic? Because *her* children brought this important issue to her attention.

According to reporting by Reuters in Amsterdam, the Dutch education minister Robbert Dijkgraaf stated that as of January 1, 2024, mobile phones, tablets, and smartwatches will be banned from classrooms. The decision is based on the scientific research showing that mobile phones are a disturbance to concentration.[318]

Until regulations are changed and industry comes up with improved and safer technologies, parents are in charge. No parent wants to be told that their action or inaction may not be right for their children. Yet, giving a child a powerful and potentially addictive pharmaceutical for a preventable and reversible form of depression or anxiety should be a last resort. If you have a child on medication for depression or anxiety, discuss the potential impact EMF may be having with your child's doctors. More than likely, your pediatrician and child psychiatrist will not know anything about the association of EMFs with anxiety and depression. But by bringing up this question, you will introduce your child's doctor to the idea, and perhaps they will research on their own. Hopefully, they will be grateful and it will help them treat the multitude of others with the same condition. Make a reference to the EMF medical conference available online.

At the same time, remove or at least reduce your child's exposure to RF/MW radiation as much as possible. In addition, bring your child into a park or woods far away from outdoor antennas a few times a week so they can ground and resync with the natural world. Perhaps after a month or so, with the guidance of your pediatrician or child psychiatrist, there may be an opportunity to lower or discontinue your child's medication. As parent groups and the medical system become aware of the impacts of EMF on children's health, hopefully politicians and industry will shift and ultimately respond.

CHAPTER 15

CONNECTING THE DOTS

Are you connecting the dots? Do you now understand that electromagnetic energy can be toxic? Is it clear there is a difference between a beneficial source of energy like the Earth's natural frequencies and toxic manmade radiation and that the latter can have harmful health effects? That small changes to the radiation frequency or intensity can be the only difference between the two?

Our present-day regulations ensure technology-emitted radiation is not high-enough frequency to cause damage through ionization. However, it has become clear over time that our daily exposures to nonionizing radiation are causing stress reactions in our cells, which can lead to negative health outcomes. Tens of thousands of research studies performed throughout many decades have documented these effects. But some physicists and engineers still claim all these peer-reviewed studies and the conclusions of their authors are erroneous *because the only possible biological effect from nonionizing radiation is heat generation.* Hopefully, the fundamental flaw in this argument is now obvious to you after reading this book.

Where Are the Regulators?

Many national and international work groups and regulating agencies have been tasked with determining the safety of nonionizing radiation, including ELF-EMF and wireless technology. These groups are interrelated to some degree, and it has become apparent in the US, in Western Europe, and in many other parts of the world that regulators have been in bed with industry at the expense of public health. The International Commission on Non-Ionizing Radiation Protection (ICNIRP), the World Health Organization (WHO), and the Federal Communications Commission (FCC) in the US are three of the major influencers that have decided the fate of the world's population with regard to radiation exposure. In 1998, ICNIRP published their initial set of guidelines, which became the template for international regulations. The science behind their policy was based on the premise that thermal effects are the only potentially harmful biological effects of RF-EMF. In 2020, ICNIRP revised their 1998 guidelines regarding human exposure to EMFs and included frequencies from 100 kHz to 300 GHz.

Given regulators are only concerned about heat generation, quantifying specific absorption rate (SAR), as initially described by Herman Schwan in 1953, remains the only method used by industry to determine whether the exposure intensity of their products is within allowable limits as defined by regulating bodies. Determination of safe exposure limits has not changed since then, despite the extensive volume of research performed since the early 1970s documenting *nonthermal effects from nonionizing radiation*. Certainly, preventing significant tissue heating from MW exposure is an important parameter that needs to be regulated to prevent damage to human health. But I have not dedicated any significant text to SAR because I feel it has been a ruse that has taken attention away from the equally, if not more, important studies documenting nonthermal effects of radiation described in the previous chapters.

Exposure limits vary from country to country (Figure 15.1). The US, Canada, and most western European nations have adopted the highest exposure limit of 10 W/m², or 10,000,000 µW/m². Several epidemiological studies have demonstrated that higher exposure limits have affected public health. A paper by a Russian scientist, Vladimir Mordachev, demonstrated a strong correlation between a country's RF/MW radiation exposure limits and the total number of deaths that country suffered early during the COVID-19 pandemic.[319] Increased mortality was also observed in US states and cities with 5G networks during the same period by Canadian researchers.[320]

Country	Max Exposure Limit (W/m²)	Country	Max Exposure Limit (W/m²)
Belgium	10	**Kazakhstan**	0.1
Bulgaria	0.01	**Lithuania**	0.1
Canada	4	**Luxemburg**	0.45
Chile	0.1	**Netherlands**	10
Denmark	10	**Poland**	0.1
France	10	**Portugal**	10
Germany	10	**Russia**	0.1
UK	10	**Spain**	10
Hungary	0.1	**Sweden**	10
India	0.9	**Switzerland**	0.1
Ireland	10	**Turkey**	0.56
Israel	0.9	**Ukraine**	0.1
Italy	0.1	**USA**	10
Japan	10	**Uzbekistan**	0.025

Table 15.1. Maximum exposure limits as of 2020. Data from Mordachev, 2020.

Russia, the country that has perhaps studied the effects of MW radiation more than any other, has a maximum exposure limit of 0.1, or 100,000 µW/m², which is *one-hundredth* of what we allow here in the US. In Bulgaria, the maximum allowable exposure is even lower, at 10,000 µW/m², a value lower than the background radiation intensity in many homes with a Wi-Fi router. In 2020, ICNIRP published updated guidelines, but they are still based on thermal effects alone and therefore dangerously misleading as they ignore the nonthermal biological effects of RF/MW radiation.

Unfortunately, as with other governmental agencies, industry has penetrated the regulatory agencies and corrupted them. The revolving door between industry and government should be well known to all of us by now. If not, see Harvard's Captured Agency report.[321] Within the World Health Organization (WHO), there are two groups studying the health effects of RF, one of which is heavily biased toward industry. But the other is an independent research group called the International Agency for Research on Cancer (IARC), which has been tasked with assessing the risk a person has for developing cancer when they are exposed to various environmental toxins. From what I've read, this group has been properly vetted to ensure their team members have no conflicts of interest with industry. This group divides agents they study into one of four categories:

- Group 1: Known carcinogen to humans
- Group 2A: Probable carcinogen to humans
- Group 2B: Possible human carcinogen
- Group 3: Not classified

In 2011, this group classified RF/MW radiation emitted by cell phones, referred to as "wireless radiation," as a Group 2B carcinogen. This is the same category as lead, thalidomide, DDT, engine exhaust, aspartame, sodium saccharin, and others. This assessment was based on a review of human epidemiological studies and animal studies published in the literature up to that time. Since then, research has continued and technology has become much more pervasive, as we've discussed. In response to the numerous scientific papers further clarifying the potential for RF/MW radiation to cause cancer, the IARC committee recommended a reevaluation of their Group 2B designation for ELF-EMF and wireless radiation in 2024, with the potential of bumping it up to a Group 2A or possibly a Group 1.

As has probably become apparent, the science behind the health effects of nonionizing radiation is complicated and likely beyond the current understanding of most politicians. In the US, the evaluation of MW radiation was initially buried in layers of bureaucracy. In response to concerns about UV radiation emission from color TVs and MW radiation from microwaves, the Radiation Control for Health and Safety Act was approved in 1968 to create an agency tasked with protecting the public from radiation emitted from electronic products. This act authorized the US Food and Drug Administration (FDA) to set federal standards and to conduct research. The FDA gave the responsibility for regulation of EMR emission and establishment of performance standards for electronic products to the Center for Devices and Radiological Health. In tandem, the Federal Communications Commission (FCC) was tasked with evaluating the effect of emissions from FCC-regulated transmitters on the quality of the human environment by the National Environmental Policy Act of 1969.

In the US, the regulating body for wireless communication is the FCC. This agency has for all intents and purposes given a green light to wireless companies, enforcing the Telecommunications Act of 1996, a law that was enacted decades ago, before deployment all of the technology we are now using. One of the most distressing clauses of this act, Section 704, specifies that the placement of cell towers and antennas cannot be regulated by communities based on environmental effects of radio frequency emission. This has been interpreted by some US circuit courts to mean that health concerns cannot be used to restrict cell tower placement. If the radiation emitting from a tower complies with FCC regulations, it is presumed to be safe. A community can only prevent placement of a cell tower if it doesn't create a gap in network coverage for a cell phone call, a severe limitation on the community's ability to care for its citizens.

The FCC is an independent agency in the US, but its members are political appointees. As we know, politicians are beholden to their campaign donors, and the telecom and wireless tech industries have enormous wealth. Industry captured the FCC years ago. Operatives muddied the research, controlled the messaging, and created a disinformation campaign replete with propaganda, touting biased industry-funded scientific studies, which confused the public and misled governmental leaders. In November 2019, the FCC failed to incorporate relevant research results from hundreds if not thousands of scientific papers that had been published since their original regulation standards. Two powerful lobbying groups, the Environmental Health Trust and Children's Health Defense, as well as some smaller groups, sued the FCC for not updating their safety standards to reflect important research findings. In December 2019, a historic ruling by the United States Court of Appeals sided with the plaintiffs acknowledging that the FCC had been "arbitrary and capricious" in not updating their 1996 safety limits.[322]

In its ruling, the Court specified that by not updating their regulations, the FCC failed to "record evidence that exposure to RF radiation at levels below the commission's current limits may cause negative health effects, including those *unrelated to cancer*." I was happy to read that this court referenced noncarcinogenic effects of RF/MW radiation in addition to environmental impacts because, as we've covered, they are contributing to the deterioration of public health in general and, as a result, increasing the cost of healthcare nationwide.

A group of 430 international scientists and physicians, known as the *5G Appeal*, have called for an immediate moratorium on all wireless expansion until more scientific experiments are performed to critically determine the safety of this technology. Many grassroots organizations have formed in the US and around the world to combat the continued roll out of this technology. If you are interested in getting your questions answered or becoming an activist, do some digging to find the group closest to you. In the US, Americans for Responsible Technology not only has a list of organizations, they also have a link to start a group and a toolkit that enables citizens to educate their fellow citizens and legislators. The Environmental Health Trust is a wonderful organization that also provides a large number of resources, including links to research papers and a tremendous number of educational materials. Other organizations also exist, but these four are a good place to start:

- Environmental Health Trust (ehtrust.org)
- Americans for Responsible Technology (www.americansforresponsibletech.org)
- Children's Health Defense (childrenshealthdefense.org)
- TechSafe Schools (techsafeschools.org)

Where Are We Heading?

We are currently on a dangerous global trajectory. Not only are 5G cells being installed all over, at close range, tens of thousands of satellites are planned to be deployed to blanket the Earth with electromagnetic radiation and become part of the 5G network. Higher MMW frequencies up to 100 GHz have been reserved for 5G expansion. And 6G is already under development at Northeastern University's Institute for the Wireless Internet of Things through support by industry and the US Department of Defense. This network will employ *even higher* frequencies, in the terahertz range (300-3,000 GHz). How this frequency will be used and for what purpose, aside from transferring data at incredibly fast speeds, is uncertain.

If industry continues to increase the intensity and frequency of emissions, we will continue to augment our exposure to EMR with unknown consequences. Exposing nearly the entire human, plant, and animal population to modulated microwave radiation frequencies, and now millimeter waves, has been one big–and dangerous!–experiment. We have all been the subjects in this protocol and, from what I can see, there is no control group. Our nerves are becoming fried and, at the same time, fatigue is widespread.

Competing technologies are in play. There are those who prefer fiber optics for information transmission, as it is faster than Wi-Fi. FiWi is an acronym for a technology that combines the use of fiber optic "wired" connections and Wi-Fi. Hardwired networks utilizing fiber optics are faster than Wi-Fi. University College London developed the hardware for a fiber-optic technology that transmits data within pulsations of light at over 1 terabyte/sec. There is a slogan in the audio-video world: "Wires for things that don't move, wireless for things that do." If we reserved wireless technology for when we are on the move, we would decrease our daily exposure significantly. Everything in the house, aside from a cell phone, would be hardwired and cell phones can be forwarded to a landline. Our children will ultimately decide the fate of this technology. Their fascination with machines may become more utilitarian as they grow up and long for connection to real people and the natural world.

Pro-tech/Anti-tech

Having written this book, I may come across as against all technology. But I believe in balance. My source of strength comes from my connection to the natural world, and I appreciate the daily wonder and sense of presence I experience when immersed in it. Through observation, the natural world provides me with opportunities to learn life's lessons. I believe we can all better understand who we are by appreciating the consciousness of the natural world.

But that appreciation doesn't make up all of "me." I also enjoy exploring new technology. As I mentioned earlier in this book, I was the first kid I knew who owned a personal computer in the 1970s. And today, I drive an electric car, which I acknowledge exposes me to high RF and magnetic fields. Why do I expose myself to high EMF knowing what I know? Even though I feel like I've been exposed to uranium for hours after going on a long drive, I like to experience technology. And, I believe that in an effort to better understand the health effects of technology, it is important for me to experience what others do. In order to protect my health, I sleep in an EMF-free environment and take a number of antioxidants to try and offset the excess oxidative stress my body undergoes each day. Then, every day, I hope and pray for good health.

The radiation coming from outdoor antennas, orbiting satellites, cell phones, and all other sources, including tracking devices, is likely affecting our health and the health of our pets and outdoor wildlife, including insects and birds.[323, 324] In addition, I suspect this radiation is having additional unexpected effects, including impacting some of humanity's most precious artwork. I have had the privilege to stand in the Sistine Chapel five different times over the past thirty-five years. The first time I went, they were just starting to restore the work, and the colors revealed during the restoration process were shockingly vibrant and fanciful. Over the past few years, I see a dinginess to the paint that I believe is happening much more quickly than it should. Strict rules enforced by the Vatican prohibit flash photography because they are trying to preserve the piece and protect it from the oxidation that can occur from light, which as we know is a form of EMR. Light fades pigment by causing oxidation. Unfortunately, I suspect the RF/MW radiation emitted by cell phones, wireless access points, and wireless security systems is oxidizing the paint pigments in the Sistine Chapel, as well as in museums all over the world, except perhaps those pieces protected by reflective glass. Museums should require all cell phones be placed in airplane mode before one is allowed to walk through and observe treasures created by the world's greatest artists.

Paradigm Shift

There are many researchers and lecturers out there providing knowledge and research experience, helping to educate others in this complex topic. One of the most enjoyable lectures I've heard on EMFs was given by the Canadian research scientist Dr. Magda Havas, a now-retired professor emeritus from Trent University in Toronto. Dr. Havas has written extensively on the subject, and many of her papers have been referenced in this book. I was fortunate enough to hear her give an entertaining and thought-provoking talk about the health effects of RF radiation to a group of healthcare providers. Although the material she presented was certainly disturbing, her warm demeaner and radiance inspired and created optimism. During her lecture, Dr. Havas described the need for a *paradigm shift* away from the biomedical model of life to one that understands there is a fundamental flow of energy guiding the biochemistry that is controlling it all. I took this statement to heart and began my journey into further understanding this complex concept.

I believe her assessment is true. Our bodies function via the organization and manipulation of energy. The movement of electrons and other charged particles inside and outside cells creates an ever-changing three-dimensional web of electromagnetic fields. Furthermore, there is an energetic coherence that establishes you as a unique being and an independent organism. RF/MW radiation can distort and weaken this coherence.

You have your own electromagnetic ID, as it were. These physiological currents are in the ELF-EMF range and radiate from your body at the speed of light. Your composite modulated frequency distinguishes you from every other organism on the planet. Recently, EMR transmission was documented as a method of communication among organisms, in this case, bacteria.[325] I suspect *all life-forms* send out encoded messages by the emission of EMR. After these signals are received, they are subconsciously responded to.

Tina: An Introduction to the Amazing Realm of Inherent Energy and Information Transmission

When was the last time you "read your partner's mind"? Tina and I were never lovers. When our experiments began, we weren't even established friends. We had met one week earlier in a school and were, in a sense, fellow explorers on a similar journey to better understand how our minds work.

In order to learn to recognize each other, Tina and I started our discipline, as we had been taught, by sitting in a lotus position on the floor, face-to-face while holding hands and staring into each other's eyes. It was a beautiful exchange that brought both of us into a trance. Energy currents flowed through our connected hands, while at the same time, ELF-EMF connections were being made by the brain. The imagery presented to our brains through our eyes was being hardwired in our minds with the complex of ELF-EMF frequencies that represented the other.

After this intimate engagement, we began a weekly exercise. At a random time during the week, we would go into a trance and focus on the other by visualizing their face with our mind. When we could clearly picture the other person, we would bring up the image of an object in our mind and try to send it to the other person telepathically. The test was to then see if the other person, when they got around to doing their own focusing, could pick up on what the other was thinking.

At first, we chose categories to narrow down the possibilities. For example, one week, one person might send the other two colored shapes, such as a yellow circle or a blue cone. The other person might send two objects found in the kitchen. This naturally increased our chances of success and built up a little confidence. Each week, we recorded hits, near misses, and total misses. If, for example, I pictured a red square, and she thought I had pictured a red circle, that was a near miss. If I thought of Thanksgiving as my favorite holiday, and she said Thanksgiving, that was a direct hit! Over the course of a few months, our accuracy palpably increased.

Then, one week, I was confronted with the extraordinary.

One Sunday, after going through our results from the previous week, Tina told me she would send me two animals over the next week. I replied by saying I would send her two playing cards. Jokers were fair game. We hung up the phone, agreeing to speak again the following weekend.

On Thursday, while driving home from work later in the week, I realized that I hadn't done the exercise yet. I began to think about what card I would choose to send to Tina. The first card that popped into my mind was the ace of spades. I chuckled to myself, thinking that wasn't the nicest card in the deck to send. When I got home, I took out a deck of cards and went through them quickly, deciding I would send Tina the two of hearts and the ten of diamonds. After writing them down in my journal, I studied the cards carefully and went through the ritual. I focused on Tina's face and then held the image of the cards with my mind. After I was done, it was time for me to receive what she had sent (or would send).

I cleared my mind, brought up her face and waited for animals to come into my mind.

The first image that came was faint. I recognized it was an animal with a long neck. *Ahh, a giraffe*, I thought. I cleared my mind again and waited for the next image to appear. The second animal I saw was in a tree. A koala bear? I wasn't sure, but I entered my observations into the journal and promptly forgot the session.

That weekend, I flew to Las Vegas for a medical conference. Sunday morning, I went down to the casino lobby and waited in line for a table at the coffee shop. It was 11:00 a.m., and time to call Tina to discuss our results.

The Results

After exchanging some pleasantries, we got down to business.

"I'll go first" she said.

"What did you get from me?" I asked.

"Well, I brought up your face while I was doing a meditative walk and the card that popped up into my mind was the ace of spades."

I gave a half laugh with a little bit of surprise.

"I'll give you credit for that!" I said. "I had thought about sending you the ace of spades but then changed my mind."

After she berated me for changing my mind, something that was against our established rules, I asked her if she got another card from me.

"Well yeah. I did. I got two more cards from you."

The hair on the back of my neck stood up. She continued,

"The next cards I saw were the two of hearts and the ten of diamonds."

I think I stopped breathing. I gulped and tears welled up in my eyes. I stood there barely able to hold on to the phone. The contrast between the extraordinary experience we had just shared and the noise and commotion of the casino was all the more disorienting.

"Oh my God!" I said to her, "That's exactly what I sent to you!"

"OK cool. What did you get from me?"

Tina was unemotional and unmoved by her accuracy. She knew it wasn't her conscious mind that figured it out, as if it were some sort of puzzle she had thoughtfully solved. She had merely accepted the information that came to her and recorded it. In her world, she expected this outcome, for we had been practicing . . . right? I, on the other hand, was dumbfounded.

Our conversation continued.

"The first thing I received from you had a long neck. At first, I thought it was a giraffe, but then I wasn't sure if it was a tortoise."

"Well, what was it?" She asked somewhat impatiently.

"A giraffe," I said.

"Yes, that's correct. I sent you a giraffe. What else did you get?"

I told her I thought the second animal was a koala bear. But that was a total miss. She had sent me something completely different.

Now, I knew this accuracy was not random or about odds. Regardless, I felt compelled to calculate the probability of her being able to accurately identify the three playing cards alone, out of a full deck of possibilities plus the jokers, that I had sent her with my mind. I came up with odds equal to 1 out of ($54 \times 53 \times 52$), or 1 out of 148,824! And, that doesn't take into account that fact that there were only supposed to be two cards sent! When considering that there were actually five bits of information going back and forth between the two of us, and we hit four of them spot-on was unbelievable and unforgettable, to say the least. It built acceptance in me not only for this one instance but also for the idea that we all do this all the time.

Are We Always Reading Each Other's Thoughts?

Tina and I continued our exploration into this work and continued to have remarkable success. Then, one day, she called me up out of the blue. "Were you looking at a red fire hydrant today? I had this image of a fire hydrant pop up in my mind, and I didn't know where it came from. I thought it was probably coming from you."

She didn't know it, but I had been house hunting that day and, at a property I really liked, I spent time looking at a red fire hydrant placed on the front corner of the property, trying to decide whether its location bothered me or not. What was so intriguing about this was that she didn't know the image was coming from me, but she had received it. Does this happen all of the time? Does it happen with people we don't necessarily even know that well?

Enormous Implications

The experiments I ran with Tina were extraordinary and, although some of the communication could theoretically be explained by exchanging thoughts via ELF-EMF, this seemed unlikely. A mysterious component of our practice was that we didn't synchronize our exchanges—they didn't happen at the same time. When I sent her the three playing cards, I focused on her on Thursday at 6PM. She might have done her discipline to receive the information from me on Friday—or even perhaps on the Tuesday before, which opens up a whole other realm of possibility.

Regardless of how deep down the rabbit hole you want to go, perhaps it is enough to know that when you look into the eyes of another, your brain is associating an image with a complex energetic signature. If you receive thoughts from this person when you are not with them, or perhaps when you are with them but words aren't being spoken, you may actually be receiving and decoding electromagnetic energy being transmitted from their brain. We long for connection—and we always have it! An intriguing podcast called *The Telepathy Tapes* delves deeply into this fascinating phenomenon with nonspeaking autistic individuals. I strongly recommend checking it out. It is clear to me that we communicate with one another not only through written and oral communication, but chemically, through the production of pheromones and other biochemicals, and perhaps most importantly, energetically.

I've heard some say there are no secrets in the spiritual world. If we do, in fact, have access to download information from those we are connected with, it is perhaps true that our thoughts and observations are shared with others. This might seem like an incredible violation of privacy for those who live in fear and choose to live a life of deception. But the flip side is that we are all connected, and that is a beautiful thing. If you live a life of truth and honest pursuit, your experience will be pure and those around you will feel a genuine connection to you. I believe electropollution is in a sense, static, making it more difficult to tune in the subtle frequencies that connect us with others, causing an intangible sense of isolation and loneliness. Perhaps the most damaging aspect of ubiquitous RF and MW radiation is that it *seems* to be connecting us, when in reality it is doing exactly the opposite.

Take time to contemplate the contents of this book and the significance of this research, not only to remediate your environment and reduce your exposure to excessive forms of electropollution, but to also understand who you are and what your body and the bodies of those around you are. We have a lot of the pieces to the puzzle.

Connect and Have a Date with Planet Earth

So how do you start? First of all, go outside. Like in the 1976 movie *Logan's Run*, it wasn't until the main characters got outside that they realized how screwed up their civilization had become. We all need to be connected to the Earth and its frequencies as often as possible. Those who work the soil are grounded and understand the strength gained by digging in the dirt, the wisdom learned by truly understanding the balance of nature and how the elements and life create balance. We learn and mature from such interactions. For those who fear the natural world or at least elements of nature, do what it takes to conquer your fears. Be a master of your domain and try surrendering to become one with your world without escaping into a synthetic digitized model of reality.

Perhaps this is one of the allures of technology. We can have a detached, safe experience and turn it off when we are done. But the natural world is not something to evolve beyond, exploit, or avoid. It is a gift, if we are wise enough to see it and experience it, if only for a few years.

We are connected without technology. If we can get comfortable turning off our devices and tuning into the natural world, our minds will begin to expand and our civilization will be the better for it. If we are lucky enough, we will go back to days with less depression and anxiety, better sleep, and improved health.

APPENDIX A

WAVE THEORY

Electromagnetic energy waves are mostly invisible, but are often represented as sine waves, as illustrated in Figure A.1.

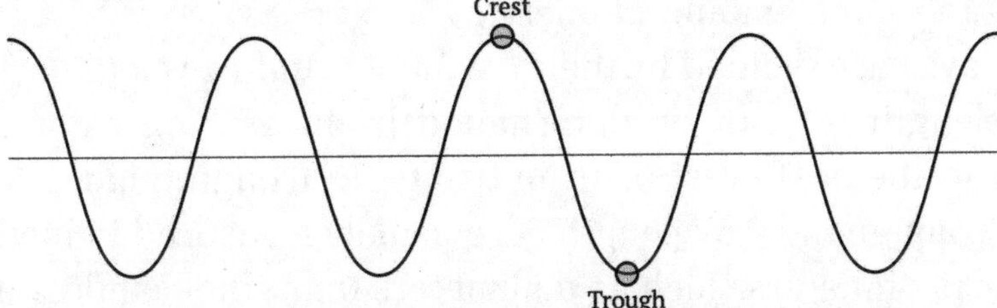

Figure 15.1 Sine wave

If you took geometry in high school, this no doubt looks familiar. The top of the wave is referred to as the crest and the bottom as the trough. The higher the crest (and lower the trough), the greater the amplitude of the wave (Figure A.2).

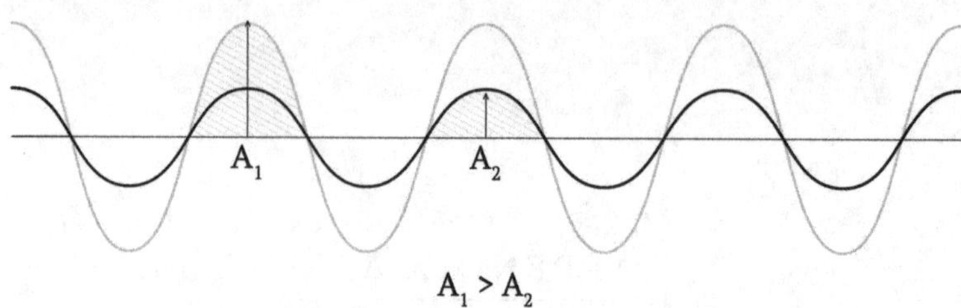

Figure 15.2. Amplitude

An increase in amplitude indicates the wave has a greater intensity and is effectively transmitting more energy. In the field of radiation physics, the term *power density* is used as a measure of that energy, because a higher intensity of radiation really means an increase in the number of photons, or packets of energy. Think of the difference in light intensity produced by a bulb on a dimmer switch: low-level light versus bright light. Bright light emits a higher number of photons, each carrying the same amount of energy.

Waves are defined by their *frequency* and *wavelength*. The wavelength (λ) can be determined by measuring from one crest to the next crest or from trough to trough (Figure A.3). Electromagnetic wavelengths are usually measured in meters or parts of meters, such as millimeters (*mm*–thousandths of a meter), micrometers (μm–millionths of a meter, also known as "microns"), and nanometers (*nm*–billionths of a meter).

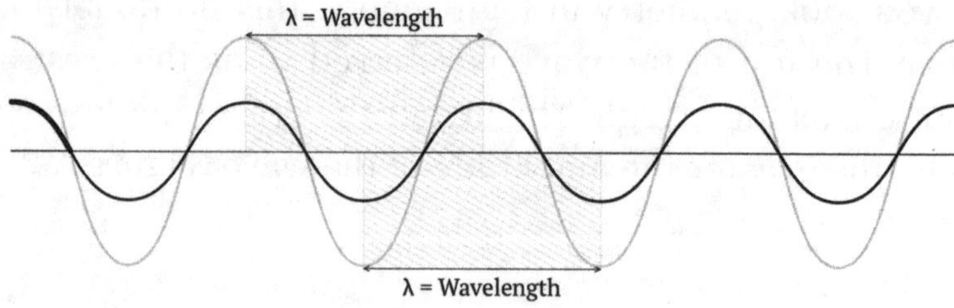

Figure 15.3. Wavelength

Electromagnetic waves irradiate from their source and travel at a constant speed, determined by the medium through which the energy is traveling. You can imagine that if a wave is moving at a certain speed and you were watching it go by, you would be able to count the number of waves that pass you in a given amount of time, like sitting in a car at a railroad intersection and counting the number of cars that go by. The number of waves that pass by in one second is referred to as the wave's frequency (v). Frequency was originally reported as *cycles per second*. But in 1960, the unit Hertz (Hz) was adopted by the General Conference on Weights and Measures to memorialize the scientist who conclusively proved the existence of electromagnetic waves, which will soon be explained.

Properties of Waves

Waves have fundamentally different characteristics than particles. For one thing, waves can more easily penetrate objects. Sound waves, for example, can go through a glass window, but a metallic BB cannot—at least, not without breaking it. Waves can also *reflect*, *diffract*, *refract*, *scatter*, be absorbed, cause *resonance*, and create *entrainment*.

After its creation, light radiates in all directions from its source, but what happens to this energy is determined by what the light encounters along its way. As mentioned, the speed of light is constant, as long as the energy waves travel through a homogeneous medium, such as air. Light travels through air with a velocity of approximately 300,000 km/sec. In water, the speed of light drops to 225,000 km/sec. As the wave travels from air to water, it therefore slows down and, as a result, changes direction. This change in direction when light encounters a new medium is an important characteristic of light energy and is called *refraction*. When we are able to observe the light in both media, we are presented with an optical illusion, like the one in Figure A.4.

Figure 15.4. Refraction of light waves

The light's wavelength does not change when it refracts. Rather, the frequency changes when the wave's velocity increases or decreases. In the above example, the frequency of light decreases as it enters the water because the speed of light in water is slower.

Reflection is another characteristic of waves. Reflection occurs when waves bounce off the surface of an object and travel in a new direction. We observe the effects of reflection daily as most of what we "see" is a result of light reflecting off object surfaces. An analogy to reflection in sound is the echo. If you stand in front of a canyon and yell, you will hear an echo of your yell coming back to you. The echo is due to the sound waves reflecting off a distant object, in this case a canyon wall, and then traveling back to you for your ears to hear.

Electromagnetic waves can also pass through solid objects, a process called *transmission*. Transmission also occurs with sound waves. Anyone who has lived in an apartment or townhouse and shared a common wall with a noisy neighbor has no doubt experienced this phenomenon. Sounds, particularly lower-frequency bass notes, travel through walls, ceilings, floors, and any other structure, causing reverberation. Light waves are similar. Longer wavelengths within the electromagnetic spectrum coming from the sun will penetrate and transmit to deeper depths than shorter wavelengths.

One of the more fascinating properties of waves is a phenomenon called *resonance*. This is the ability for the energy in a wave to become absorbed by a physical structure. Resonance can cause the object to vibrate and generate heat. Resonance indicates a special type of relationship that exists between the energy source and energy receiver. In order for resonance to occur, an energy source must produce repeated waves of a single frequency and wavelength, such as the sound waves produced by playing a tonal note.

Periodicity is the characteristic of wave repetition. *Noise*, on the other hand, does not demonstrate periodicity and cannot contribute to resonance. All objects resonate at a specific frequency, which is determined by each object's shape, length, thickness, and composition characteristics. An object's resonant frequency is the frequency that it maximally absorbs. Even tiny structures like molecules and atoms have resonant frequencies. In fact, some of our most advanced technology in medicine is based on this phenomenon, such as magnetic *resonance* imaging (MRI). In some circles, people say they are "frequency specific" with one another. This means they feel connected to one another and share similar ideas and energy. In other words, people who are frequency specific resonate with each other.

Resonance is simple to demonstrate. If you place two stringed instruments next to each other, such as a violin and a guitar, playing a note on one will cause a corresponding string on the other to vibrate with the same frequency. For example, if you place your index finger in first position on the first violin string and stroke the string with a bow, the tone played will be the note "A." Perhaps to your surprise, the second string from the top of a conventionally tuned guitar will vibrate and produce the same sound. Why? Although the guitar string is much longer than the violin string, each string is of a specific thickness and material that allows it to vibrate at the same frequency, which is 110 times a second (110 Hz). Because of resonance, the guitar string will absorb the sound waves coming from the violin, take on their energy, and vibrate, producing its own sound.

The cliché of the opera singer sustaining a high note and shattering a wineglass is another example of resonance. Depending on its contents, shape, size, and crystal composition, a wineglass also has a specific resonant frequency. If the opera singer can produce waves of that frequency by singing the note that corresponds to it, the glass will absorb the sound waves, vibrate, and even perhaps shatter if the intensity of the waves is great enough to overcome the binding strength of the crystal.

You may or may not remember that the old Tacoma Narrows Bridge, referred to as "Galloping Gertie," collapsed in 1940 shortly after it was opened. The bridge's failure was originally believed to be due to resonance, assuming that during a storm, the bridge was subjected to the wind's energy blowing at the same frequency as the bridge's resonant frequency. The bridge would have absorbed energy from the wind and begun to sway and bend, leading to its demise. It turns out that a different physical phenomenon, "fluttering," was the true culprit, but bridge resonance does occur and such a collapse is theoretically possible. Because light is transmitted through waves, resonance can and does occur with this form of radiation too.

Another fascinating feature of cyclical energy is its capability to cause entrainment. If you have ever walk into a cuckoo clock store, you may notice that all of the pendula were swing at the same speed and in synchrony with one another, meaning they all swing in the same direction at the same time. I once thought that the shop owner had carefully manipulated the pendula to get this to occur, kind of as a "wow" effect for the potential customer walking into his shop. What I didn't know was that this is a natural phenomenon that occurs with cyclical or repetitive systems, including waves.

Our organ systems entrain with one another. When you consciously slow down your breathing rate, your heart rate will also slow and the chatter in your brain will quiet down. This is a method many people use to enter a state of meditation. Slow, deep breaths cause a relaxed meditative state through entrainment. Listening to slow music will slow down your body systems and help you relax. Alternatively, when you listen to music with a fast beat, you will become energized and, in a sense, "speed up." Living systems can entrain to music as well as to sources of light and other forms of radiation. The creation of circadian rhythms is an important example of entrainment to periodic cycles of light and dark.

There is much more to know about the physics of waves, but these definitions and concepts should help you better grasp the basic physics presented throughout this book.

SAMPLE WORKSHEET FOR SURVEYING ELECTRIC FIELDS

Bedroom

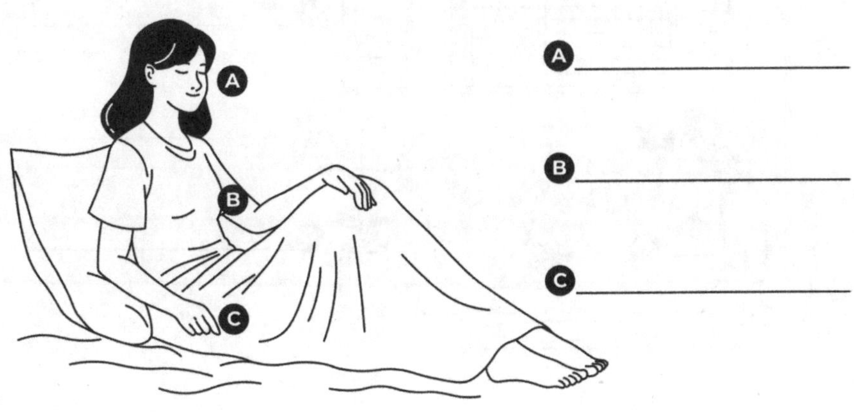

Office

A ________________

B ________________

C ________________

Living Room/Den

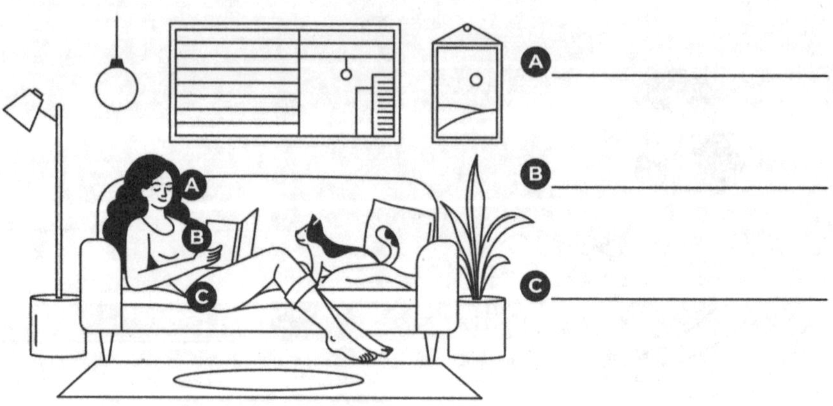

A ________________

B ________________

C ________________

SAMPLE WORKSHEET FOR SURVEYING MAGNETIC FIELDS

Bedroom

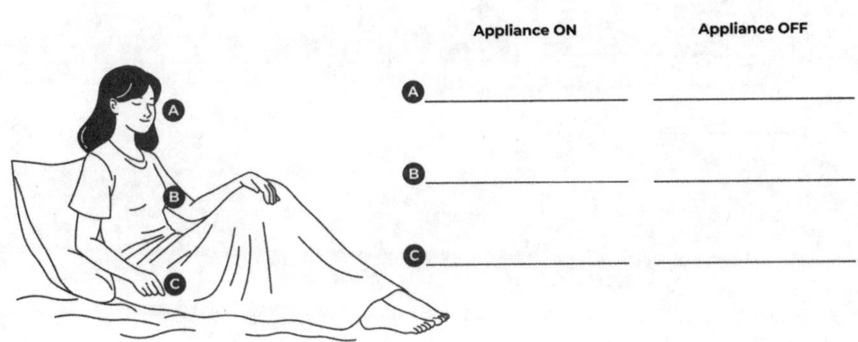

Office

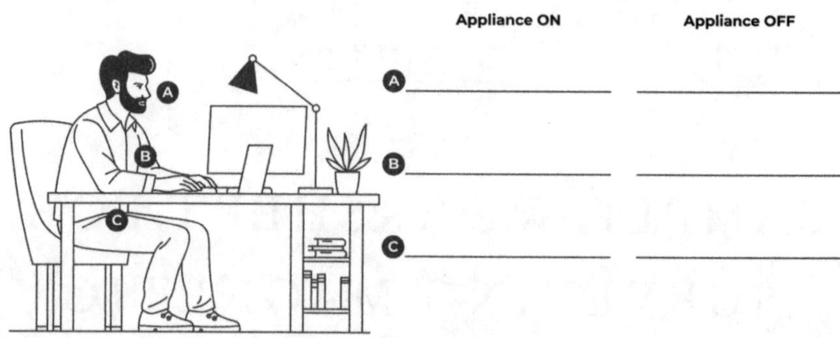

Living Room/Den

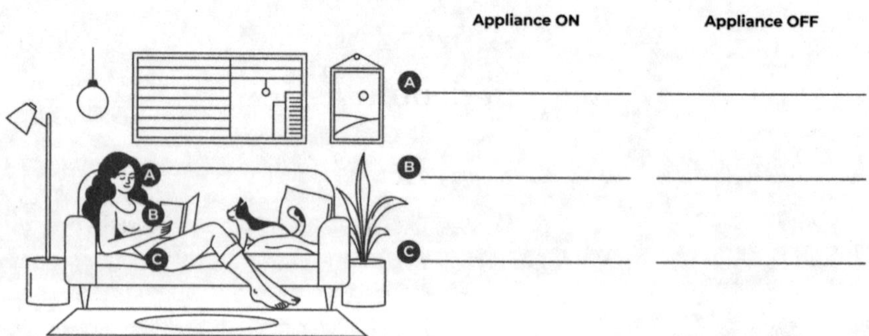

SAMPLE WORKSHEET FOR SURVEYING MICROWAVES AND MILLIMETER WAVES IN HOME

Device	Distance	Intensity
Router		
DECT phone		
Wireless speaker		
Security system		
Baby monitor		
Cell phone		

ACKNOWLEDGMENTS

Putting together a manuscript is a team effort, and many friends and colleagues have helped to modify and augment this work. I would like to take a moment to thank them all for their contributions. Seth Wolpert, my brother-in-law, and Kent Chamberlin, both engineering professors, helped edit and skillfully explained some of the more abstract engineering concepts and related content. Thank you, Diane Vadino, for your stylistic guidance in helping to aggregate the most suitable artwork and charts, and Becky Wolpert and Natalilia Koh for providing original artwork.

The first read-through was by my lifelong friend, David Itkin, who provided preliminary edits and encouragement. A trailblazer in this field, Cece Doucette, the president of Massachusetts for Safe Technology, provided expert editorial guidance and helpful recommendations. My dear friend and advocate, Paul Mavrinac, who I think has read most everything I've written, studied this manuscript and contributed important perspectives and editorial input. Paul's belief in the importance of this material and his unyielding support and encouragement have been a tremendous comfort during this process, and particularly during difficult times.

I am extremely grateful to my good friend, Patty Lemer, for her support and for recommending and then editing the glossary. Thank you, Patty, for connecting me to Lou Conte with Skyhorse Publishing, who importantly saw the value of this work. Once in the hands of Skyhorse, this manuscript received a thorough line and content edit by Zoey O'Toole, who did an outstanding job parsing through each word. I am greatly blessed to have been handed such a knowledgeable and professional editor in Zoey. Finally, I would like to thank my publication editors, Caroline, who helped select the cover design; and Carter, whose enthusiasm and guidance carried the torch to the finish line.

My heartfelt gratitude goes to Devra Davis, an epidemiologist and founder of the non-profit organization Environmental Health Trust. I originally sent Devra the manuscript looking for an endorsement. After taking a look, she sent me back a note that simply said, "We have to talk." The result has been a beautiful friendship and a proliferative partnership. Devra has been an exceptional addition to my life, and I am grateful to her for inviting me to join an outstanding diverse team of professionals, including Kent Chamberlin, Joe Sandri, and Sharon Kehnemui, at Environmental Health Trust.

My network of family and friends has been instrumental in giving me time and space to research, write, and rant. Thank you to my extraordinary children, Arthur and Ursula, my sisters, Jackie and Marj, cousins, Alan and Roberta, Elke, and Bob Siefers, for consistently being there for me along this journey, and believing in this work.

GLOSSARY

Ambient light: A term referring to the presence of background illumination within an environment.

Amplitude: A measure of a wave's height, which corresponds to its intensity.

Arrhythmia: An abnormal heart rhythm. The most common arrhythmia is atrial fibrillation, but many other types of arrhythmias exist which can be diagnosed with an *ECG*.

Attenuation: The reduction in *amplitude* of a wave as it passes through and interacts with matter, typically caused in part by absorption and scattering of energy as it passes through an object.

Bandwidth: A collection of contiguous frequencies within a designated range of the *electromagnetic spectrum*. Specific frequency ranges are typically purchased by communications companies to transmit and receive their wireless content.

Biophysics: An interdisciplinary field of scientific study that applies the principles of physics and the hard sciences of math, engineering, and chemistry to explain biological phenomena.

Cell phone radiation: The electromagnetic energy (encoded with information) emitted by a mobile phone that can be picked up by nearby cell towers for the purpose of wireless communication.

Dirty electricity: The contamination of an electrical line with higher-frequency voltage transients created by manipulating power, commonly occurring from smart meters, some appliances, and switch-mode power supplies that convert electrical current from AC to DC. Voltage transients generate sporadic electromagnetic fields in the *VLF* and *LF frequency* bands.

EHS: An acronym for *Electrohypersensitivity syndrome*.

Electrocardiogram (ECG): A tracing of the heart's electrical conduction obtained by attaching multiple wires–"leads"– to a patient's skin.

Electrohypersensitivity syndrome: A condition in which someone experiences a multitude of symptoms when exposed to *electropollution*.

Electromagnetic field: The cloud of energy waves emitted by sources of radiation. Because the electric and magnetic components of *electromagnetic radiation* can only be discriminated and measured independently at very low frequencies, this term is sometimes reserved for *extremely low frequencies* of *EMR,* but more often, researchers refer to the gamut of EMR when using this terminology.

Electromagnetic radiation: A form of energy composed of electric and magnetic waves that travel at the speed of light and share the same *frequency* and are oriented 90 degrees off axis to one another. The *frequency* and *wavelength* of these waves impact their interactions with matter and living tissues.

Electromagnetic spectrum: A continuum of *frequencies* (and corresponding *wavelengths)* of *electromagnetic radiation* that range from direct current, with a *frequency* of zero, through the highest, most intense *frequencies* of *gamma radiation.*

Electropollution: The odorless and invisible ubiquitous presence of human-generated electric and magnetic fields in the environment which can interfere with natural forms of electromagnetic radiation.

ELF: An acronym for *extremely low frequency.*

EMF: An acronym for *electromagnetic field.*

EMF detector: An instrument designed to sense the presence and measure the intensity of ELF-EMF and microwave radiation. Because the number of *frequencies* comprising the *electromagnetic spectrum* is so enormous, specific receiver technology is usually needed to survey for a specific frequency band. For example, Geiger counters designed to pick up very high frequencies of EMR cannot pick up *microwaves* or other forms of *nonionizing radiation.* This means several units may be required to get an accurate map of your home's radiation issues.

EMR: An acronym for *electromagnetic radiation.*

Energy: In the context of this book, this term refers to any *frequency* of *electromagnetic radiation* as all *EMR* can impact and cause a reaction on a physical structure.

Entrainment: The natural process of synchronization over time that occurs between objects that share a repetitive motion or cyclical rhythm.

Excitation: A term used to describe the movement of an electron into a higher orbital state after absorbing a specific amount of *energy* from *electromagnetic radiation*, typically referred to as a photon.

Extremely low frequency (ELF): The bandwidth of electromagnetic radiation between 1 Hz and 300 Hz. The most common source of technology-generated ELF comes from ordinary wiring transmitting alternating current to power electrical devices.

Faraday cage: A continuous enclosure composed of materials that reflect *electromagnetic fields*. The resulting interior environment is electromagnetically clean.

Frequency: The number of wave oscillations that pass by a stationary point within one second. This defining wave characteristic remains constant unless the wave is modulated.

Gamma rays: The highest known frequency *bandwidth,* from 10^{20} to 10^{24} Hz, with the shortest corresponding wavelengths, less than 10^{-12} m, within the *electromagnetic spectrum.*

Hertz (Hz): A standard unit of measure used to quantify cycles per second in honor of the German physicist Heinrich Hertz.

High frequency: A term often used to refer to a subset of radio frequency radiation between 3 and 30 MHz.

Infrared: A term given to the *bandwidth* of *electromagnetic radiation* below visible light with frequencies spanning from 10^{13}-10^{14} Hz and corresponding wavelengths between 25 μm-2.5 μm.

Ionizing radiation: The higher frequencies of the *electromagnetic spectrum*–far *ultraviolet* rays through *gamma rays*–which carry enough energy to potentially cause electrons to be ejected from an atom, depending on the *frequency* and how strongly the electron is bound to the atom.

Laser: An acronym for light amplification by stimulated emission of radiation. In this technology, wavelengths of light are reflected and superimposed on one another and made coherent, which amplifies the energy wave and allows it to be focused and transmitted over great distances.

LF: An acronym for low frequency.

Light: Sometimes refers to the entire *electromagnetic spectrum*, but is often used to refer to visible light–the relatively narrow band of frequencies between infrared and ultraviolet that can be detected by cells in our eyes. Thus, this bandwidth–*frequencies* from 420 to 750 terahertz, and corresponding *wavelengths* from 750 to 420 nm–is the only one that can be "seen" by the eyes.

Longitudinal wave: A type of energy transmission in which there is a to-and-fro movement of particles in the same direction as the energy's path. The most common example is sound waves, which travel through a medium–such as air–by causing the compression and rarefaction of atoms as it propagates.

Low frequency (LF): This is a term reserved for *EMFs* in the lower end of the *radiofrequency radiation bandwidth* between 30 and 300 kHz, which are often emitted by wiring contaminated by voltage transients–referred to as dirty electricity.

Melatonin: A hormone produced by the *pineal gland* in the brain during prolonged darkness that syncs the body's daily function, including the sleep-wake cycle, to periods of light and dark created by the Earth's rotation. In addition, this hormone is an important antioxidant and has many other beneficial effects on the body.

Microwaves (MW): The *bandwidth* of *electromagnetic radiation* between *radio waves* and the *infrared,* from 300 MHz to 300 GHz with corresponding wavelengths between 30 cm and 1 mm.

Millimeter waves (MMW): A subset of *electromagnetic radiation* in the higher-frequency *microwave bandwidth* between 30 GHz and 300 GHz.

Nanometer (nm): A unit of length that represents one billionth of a meter.

Nonionizing radiation: Any of the lower frequencies of the *electromagnetic spectrum* from *radio waves* through mid-energy *ultraviolet rays*, so named because these frequencies of *EMR* do not carry enough energy to cause an electron to be ejected from an atom's nucleus.

Paradigm: A term used to represent a generally accepted model or thought process to explain a scientific observation or subject matter.

Penetration: The process by which energy can pass into or through an object.

Penetration depth: A quantification of how deeply *electromagnetic radiation* passes into a biological tissue. This variable is dependent on *attenuation*. For standardization, this is the point at which the wave's *amplitude* has decreased to approximately 37 percent (1/e) of its initial height.

Photosynthesis: The process by which chlorophyll molecules within a plant convert *nonionizing radiation* in the visible *light bandwidth* into electrochemical energy by creating molecules of sugar from carbon dioxide and water and generating oxygen as a by-product.

Physics: The scientific discipline that examines the properties of particles and physical structure, in addition to radiant forms of energy including *electromagnetic fields*, sound, electricity, and magnetism.

Pineal gland: A small endocrine gland located in the center of the brain responsible for the production and release of the hormone *melatonin*.

Radiation: A term given to the emission and movement of energy waves or particles traveling in all directions from a central point of origin.

Radiation physics: An interdisciplinary science that studies the interrelationship of *radiation* and physical matter.

Radiofrequency (RF): The broad *bandwidth* of *electromagnetic radiation* between 3 kHz and 300 GHz, which is further subdivided into nine different frequency bands.

Radiologist: A physician who interprets the images produced by exposing the body to forms of energy–including electromagnetic fields, ionizing radiation, and ultrasound–for the purpose of diagnosing the presence or absence of disease.

Radio waves: The same as *radiofrequency, electromagnetic radiation* with *frequencies* less than 300 GHz and corresponding *wavelengths* greater than 1 mm. *ELF, VLF, LF,* high-frequency and extremely high-frequency bandwidths are all included within this classification.

Refraction: A phenomenon in which a wave traveling from one medium to another is deflected and changes direction.

Resonance: A phenomenon in which a structure absorbs energy, resulting in an increased *amplitude* of the structure's inherent vibration. The maximal absorption of energy occurs when the vibratory *frequency* of the energy source and the inherent *frequency* of the structure absorbing the energy are the same or similar.

Resonant frequency: This term represents the rate at which an object normally vibrates. All objects inherently vibrate.

Sine wave: A continuous S-shaped curve that represents a repetitive oscillation with a constant *amplitude* over time.

Smart meter: A device that wirelessly transmits usage patterns and energy consumption data to a utility company.

Transmission: The conveyance of electricity, radio waves, and/or information from one point to another.

Transverse wave: A type of *transmission* in which vibrations occur perpendicular to the energy's path. *Electromagnetic radiation* and ocean waves propagate with this form of *transmission*.

Ultraviolet (UV): A *bandwidth* of *electromagnetic radiation* just above visible light in the electromagnetic spectrum which includes frequencies between 10^{15} and 10^{17} Hz and corresponding wavelengths between 400 nm-1 nm. This bandwidth has been subdivided into UV-A, UV-B, and UV-C–with UV-A having the longest *wavelengths* and lowest *frequencies*.

Very low frequency (VLF): The *bandwidth* of *electromagnetic radiation* between 3 and 30 kHz. Voltage transients–also referred to as dirty electricity–are a common source of *radiation* in this *bandwidth*.

Vitamin D: An essential fat-soluble nutrient that is naturally produced by the skin when it is exposed to UV radiation in sunlight. This bioactive chemical compound, known as calcitriol, is a hormone with important roles including calcium absorption, immune function, and many others.

Wavelength: The distance between successive crests or troughs inherent in the repetitive, cyclical propagation of energy waves.

Wireless communication radiation (WCR): A term often used to describe modulated radiofrequency radiation, including microwaves and millimeter waves, used to convey information between appliances, gadgets, mobile phones, cellular networks, and related devices.

X-rays: A *bandwidth* of *electromagnetic radiation* just above *UV radiation* in the *electromagnetic spectrum,* which includes *frequencies* between 30 petahertz (10^{15} Hz) to 30 exahertz (10^{18} Hz), and corresponding *wavelengths* between 0.01 to 10 nm.

NOTES

1   S. S. Merchant, "Trace Metal Utilization in Chloroplasts," in *The Structure and Function of Plastids: Advances in Photosynthesis and Respiration*, vol. 23, ed. R. R. Wise and J. K. Hoober (Springer, 2007), https://doi.org/10.1007/978-1-4020-4061-0_10.

2   C. H. Johnson et al., "Timing the Day: What Makes Bacterial Clocks Tick?" *Nature Reviews Microbiology* 15, no. 4 (April 1, 2017): 232–42, https://doi.org/10.1038/nrmicro.2016.196.

3   R. Liu et al., "Cyanobacterial Viruses Exhibit Diurnal Rhythms During Infection," *Proceedings of the National Academy of Sciences* 116, no. 28 (2019): 14077–82, https://doi.org/10.1073/pnas.1819689116.

4   I. V. Zhdanova, "Melatonin," in *Encyclopedia of the Neurological Sciences*, 2nd ed., ed. M. J. Aminoff and R. B. Daroff (Academic Press, 2014), 1030–33, https://doi.org/10.1016/B978-0-12-385157-4.00559-5.

5   L. T. Nash, "Moonlight and Behavior in Nocturnal and Cathemeral Primates, Especially Lepilemur Leucopus: Illuminating Possible Anti-Predator Efforts," in *Developments in Primatology: Progress and Prospects*, ed. S. L. Gursky and K. A. I. Nekaris (Springer, 2007), 173–87.

6   E. Naylor, *Moonstruck: How Lunar Cycles Affect Life* (Oxford University Press, 2015).

7   T. A. Wehr, "Bipolar Mood Cycles and Lunar Tidal Cycles," *Molecular Psychiatry* 23 (2017): 923–31, https://doi.org/10.1038/mp.2016.263.

8   P. P. Posadzki et al., "Melatonin and Health: An Umbrella Review of Health Outcomes and Biological Mechanisms of Action," *BMC Medicine* 16 (2018): 18, https://doi.org/10.1186/s12916-017-1000-8.

9   K. J. Navara and R. J. Nelson, "The Dark Side of Light at Night: Physiological, Epidemiological, and Ecological Consequences," *Journal of Pineal Research* 43, no. 3 (2007): 215–24, https://doi.org/10.1111/j.1600-079X.2007.00473.x.

10 M. Figueiro and D. Overington, "Self-luminous Devices and Melatonin Suppression in Adolescents," *Lighting Research & Technology* 48, no. 8 (2016): 966–75, https://doi.org/10.1177/1477153515584979.

11   M. Hatori et al., "Global Rise of Potential Health Hazards Caused by Blue Light-Induced Circadian Disruption in Modern Aging Societies," *NPJ Aging Mechanisms of Disease* 3 (2017): 9, https://doi.org/10.1038/s41514-017-0010-2.

12   T. N. Nagarajan et al., "Blue LED Light Exposure Induces Metabolic Rewiring in Vitreous Tissues in Rat Models," *Journal of King Saud University – Science* 34, no. 4 (2022): 101986, https://doi.org/10.1016/j.jksus.2022.101986.

13   Z.-C. Zhao et al., "Research Progress About the Effect and Prevention of Blue Light on Eyes," *International Journal of Ophthalmology* 11, no. 12 (2018): 1999–2003, https://doi.org/10.18240/ijo.2018.12.20.

14   X. Ouyang et al., "Mechanisms of Blue Light-Induced Eye Hazard and Protective Measures: A Review," *Biomedicine & Pharmacotherapy* 130 (2020): 110577, https://doi.org/10.1016/j.biopha.2020.110577.

15   Nagarajan et al., "Blue LED Light Exposure," 101986.

16   C. Sanchez-Ramos et al., "Retinal Protection from LED-Backlit Screen Lights by Short Wavelength Absorption Filters," *Cells* 10, no. 11 (2021): 3248, https://doi.org/10.3390/cells10113248.

17   S. Benedetto et al., "E-readers and Visual Fatigue," *PLoS One* 8, no. 12 (2013): e83676, https://doi.org/10.1371/journal.pone.0083676.

18   A.-M. Chang et al., "Evening Use of Light-Emitting eReaders Negatively Affects Sleep, Circadian Timing, and Next-Morning Alertness," *Proceedings of the National Academy of Sciences* 112, no. 4 (2015): 1232–37, https://doi.org/10.1073/pnas.1418490112.

19   D. M. Minch et al., "Is Melatonin the 'Next Vitamin D'?: A Review of Emerging Science, Clinical Uses, Safety, and Dietary Supplements," *Nutrients* 14, no. 19 (2022): 3934, https://doi.org/10.3390/nu14193934.

20   M. F. Holick et al., "Skin as the Site of Vitamin D Synthesis and Target Tissue for 1,25-dihydroxyvitamin D3. Use of Calcitriol (1,25-dihydroxyvitamin D3) for Treatment of Psoriasis," *Archives of Dermatology* 123, no. 12 (1987): 1677–83a, PMID: 2825606.

21   A. Green et al., "Daily Sunscreen Application and Betacarotene Supplementation in Prevention of Basal-Cell and Squamous-Cell Carcinomas of the Skin: A Randomised Controlled Trial," *Lancet* 354, no. 9180 (1999): 723–29, https://doi.org/10.1016/S0140-6736(98)12168-2.

22   A. C. Green et al., "Reduced Melanoma After Regular Sunscreen Use: Randomized Trial Follow-Up," *Journal of Clinical Oncology* 29, no. 3 (2011): 257–63, https://doi.org/10.1200/JCO.2010.28.7078.

23   J. Hahn et al., "Vitamin D and Marine Omega 3 Fatty Acid Supplementation and Incident Autoimmune Diseases: VITAL Randomized Controlled Trial," *BMJ* 376 (2022): e066452, https://doi.org/10.1136/bmj-2021-066452.

24 E. Ruggiero et al., "Upconverting Nanoparticles for the Near Infrared Photoactivation of Transition Metal Complexes: New Opportunities and Challenges in Medicinal Inorganic Photochemistry," *Dalton Transactions* 45, no. 33 (2016): 13012–20, https://doi.org/10.1039/C6DT01428C.

25 G. Ablon, "Phototherapy with Light Emitting Diodes: Treating a Broad Range of Medical and Aesthetic Conditions in Dermatology," *The Journal of Clinical and Aesthetic Dermatology* 11, no. 2 (2018): 21–27, PMID: 29552272.

26 Merchant, "Trace Metal Utilization in Chloroplasts."

27 Ö Hallberg and O. Johansson, "Melanoma Incidence and Frequency Modulation (FM) Broadcasting," *Archives of Environmental Health: An International Journal* 57, no. 1 (2002): 32–40, https://doi.org/10.1080/00039890209602914.

28 O. Johansson et al., "Skin Changes in Patients Claiming to Suffer from 'Screen Dermatitis': A Two-Case Open-Field Provocation Study," *Experimental Dermatology* 3, no. 5 (1994): 234–38, https://doi.org/10.1111/j.1600-0625.1994.tb00282.x.

29 M. Havas, "Radiation from Wireless Technology Affects the Blood, Heart, and the Autonomic Nervous System," *Reviews on Environmental Health* 28, no. 2–3 (2013): 75–84, https://doi.org/10.1515/reveh-2013-0004.

30 J. M. Jbireal et al., "Disturbance in Haematological Parameters Induced by Exposure to Electromagnetic Fields," *Hematology & Transfusion International Journal* 6, no. 6 (2018): 242–51, https://doi.org/10.15406/htij.2018.06.00193.

31 B. Rubik, "Does Short-Term Exposure to Cell Phone Radiation Affect the Blood?" *Wise Traditions in Food, Farming, and the Healing Arts* 15, no. 4 (2014): 19–28, https://www.westonaprice.org/health-topics/does-short-term-exposure-to-cell-phone-radiation-affect-the-blood/.

32 R. R. Brown and B. Biebrich, "Hypothesis: Ultrasonography Can Document Dynamic in Vivo Rouleaux Formation Due to Mobile Phone Exposure," *Frontiers in Cardiovascular Medicine* 12 (2025), https://doi.org/10.3389/fcvm.2025.1499499.

33 P. Riha et al., "The Effect of Rouleaux Formation on Blood Coagulation," *Clinical Hemorheology and Microcirculation* 17, no. 4 (1997): 341–46, PMID: 9493903.

34 Liu et al., "Cyanobacterial Viruses Exhibit Diurnal Rhythms During Infection," 14077–82.

35 J. Friedman et al., "Mechanism of Short-Term ERK Activation by Electromagnetic Fields at Mobile Phone Frequencies," *Biochemical Journal* 405 (2007): 559–68, https://doi.org/10.1042/BJ20061653.

36 J. Walleczek, "Electromagnetic Field Effects on Cells of the Immune System: The Role of Calcium Signaling," *The FASEB Journal* 6, no. 13 (1992): 3177–85, https://doi.org/10.1096/fasebj.6.13.1397839.

37 M. L. Pall, "Electromagnetic Fields Act Via Activation of Voltage-Gated Calcium Channels to Produce Beneficial or Adverse Effects," *Journal of Cellular and Molecular Medicine* 17, no. 8 (2013): 958–65.

38 V. Ullrich and H.-J. Apell, "Electromagnetic Fields and Calcium Signaling by the Voltage Dependent Anion Channel," *Open Journal of Veterinary Medicine* 11, no. 1 (2021): 57–86, https://doi.org/10.4236/ojvm.2021.111004.

39 R. Munter, "Advanced Oxidation Processes-Current Status and Prospects," *Proceedings of the Estonian Academy of Sciences. Chemistry* 50, no. 2 (2001): 59–80, https://www.kirj.ee/public/va_ke/k50-2-1.pdf.

40 Y. Chen and W. G. Junger, "Measurement of Oxidative Burst in Neutrophils," *Methods in Molecular Biology* 844 (2012): 115–24, https://doi.org/10.1007/978-1-61779-527-5_8.

41 H. J. Forman and M. Torres, "Reactive Oxygen Species and Cell Signaling Respiratory Burst in Macrophage Signaling," *American Journal of Respiratory and Critical Care Medicine* 166 (2002): 1, https://doi.org/10.1164/rccm.2206007.

42 A. W. Caliri et al., "Relationships Among Smoking, Oxidative Stress, Inflammation, Macromolecular Damage, and Cancer," *Mutation Research/Reviews in Mutation Research* 787 (2021): 108365, https://doi.org/10.1016/j.mrrev.2021.108365.

43 R. O. Sule et al., "A Common Feature of Pesticides: Oxidative Stress – the Role of Oxidative Stress in Pesticide-Induced Toxicity," *Oxidative Medicine and Cellular Longevity* (2022): 5563759, https://doi.org/10.1155/2022/5563759.

44 D. G. Deavall et al., "Drug-Induced Oxidative Stress and Toxicity," *Journal of Toxicology* 2012 (2012): 645460, https://doi.org/10.1155/2012/645460.

45 Z. Fu and S. Xi, "The Effects of Heavy Metals on Human Metabolism," *Toxicology Mechanisms and Methods* 30, no. 3 (2020): 167–76, https://doi.org/10.1080/15376516.2019.1701594.

46 F. B. Teixeira et al., "Exposure to Inorganic Mercury Causes Oxidative Stress, Cell Death, and Functional Deficits in the Motor Cortex," *Frontiers in Molecular Neuroscience* 11 (2018): 125, https://doi.org/10.3389/fnmol.2018.00125.

47 E. Heyno et al., "Origin of Cadmium-Induced Reactive Oxygen Species Production: Mitochondrial Electron Transfer Versus Plasma Membrane NADPH Oxidase," *New Phytologist* 179, no. 3 (2008): 687–99, https://doi.org/10.1111/j.1469-8137.2008.02512.x.

48 E. I. Azzam et al., "Ionizing Radiation-Induced Metabolic Oxidative Stress and Prolonged Cell Injury," *Cancer Letters* 327, no. 1–2 (2012): 48–60, https://doi.org/10.1016/j.canlet.2011.12.012.

49   A. Akbari et al., "Oxidative Stress as the Underlying Biomechanism of
      Detrimental Outcomes of Ionizing and Non-ioninizing Radiation on
      Human Health: Antioxidant Protective Strategies," *Zahedan Journal of
      Research in Medical Sciences* 21, no. 4 (2019): e85655, https://doi.org
      /10.5812/zjrms.85655.

50   W. Zhang et al., "Protein Oxidation: Basic Principles and Implications for
      Meat Quality," *Critical Reviews in Food Science and Nutrition* 53, no. 11
      (2013): 1191–201, https://doi.org/10.1080/10408398.2011.577540.

51   T. Kietzmann et al., "The Epigenetic Landscape Related to Reactive
      Oxygen Species Formation in the Cardiovascular System," *British Journal
      of Pharmacology* 174, no. 12 (2017): 1533–54, https://doi.org/10.1111
      /bph.13792.

52   D. Ziech et al., "Reactive Oxygen Species (ROS)-Induced Genetic and
      Epigenetic Alterations in Human Carcinogenesis," *Mutation Research/
      Fundamental and Molecular Mechanisms of Mutagenesis* 711, no. 1–2–3
      (2011): 167–73, https://doi.org/10.1016/j.mrfmmm.2011.02.015.

53   V. Sharma et al., "Oxidative Stress at Low Levels Can Induce Clustered
      DNA Lesions Leading to NHEJ Mediated Mutations," *Oncotarget* 7, no.
      18 (2016): 25377–25390, https://doi.org/10.18632/oncotarget.8298.

54   T. Hamzehloie et al., "The Role of Tumor Protein 53 Mutations in
      Common Human Cancers and Targeting the Murine Double Minute
      2-P53 Interaction for Cancer Therapy," *Iranian Journal of Medical Sciences*
      37, no. 1 (2012): 3–8, PMID: 23115424.

55   L. Buizza et al., "Conformational Altered p53 as an Early Marker of
      Oxidative Stress in Alzheimer's Disease," *PLoS ONE* 7, no. 1 (2012):
      e29789, https://doi.org/10.1371/journal.pone.0029789.

56   S. Reuter et al., "Oxidative Stress, Inflammation, and Cancer: How Are
      They Linked?" *Free Radical Biology and Medicine* 49, no. 11 (2010):
      1603–16, https://doi.org/10.1016/j.freeradbiomed.2010.09.006.

57   V. G. Clatici et al., "Diseases of Civilization – Cancer, Diabetes, Obesity
      and Acne – the Implication of Milk, IGF-1 and mTORC1," *Maedica
      (Bucur)* 13, no. 4 (2018): 273–81, https://doi.org/10.26574
      /maedica.2018.13.4.273.

58   M. Peraica et al., "Oxidative Stress in Workers Occupationally Exposed to
      Microwave Radiation," *Toxicology Letters* 180 (2008): 38–39, https://doi
      .org/10.1016/j.toxlet.2008.06.668.

59   V. Garaj-Vrhovac et al., "Assessment of Cytogenetic Damage and
      Oxidative Stress in Personnel Occupationally Exposed to the Pulsed
      Microwave Radiation of Marine Radar Equipment," *International Journal
      of Hygiene and Environmental Health* 214 (2011): 59–65, https://doi.org
      /10.1016/j.ijheh.2010.08.003.

60   I. Yakymenko et al., "Oxidative Mechanisms of Biological Activity of Low Intensity Radiofrequency Radiation," *Electromagnetic Biology and Medicine* 35 (2016): 186–202, https://doi.org/10.3109/15368378.2015.1043557.

61   D. Hartsell et al., "The Effects of Ground Currents on Dairy Cows: A Case Study," *The Bovine Practitioner* 28 (1994): 71–78, https://doi.org/10.21423/bovine-vol1994no28p71-78.

62   D. Hillman et al., "Relationship of Electric Power Quality to Milk Production of Dairy Herds – Field Study with Literature Review," *Science of the Total Environment* 447 (2013): 500–14, https://doi.org/10.1016/j.scitotenv.2012.12.089.

63   M. Havas and D. Colling, "Wind Turbines Make Waves: Why Some Residents Near Wind Turbines Become Ill," *Bulletin of Science, Technology & Society* 31, no. 5 (2011): 414–26, https://doi.org/10.1177/0270467611417852.

64   J. Swanson, "Residential Power-Frequency Electric and Magnetic Fields: Sources and Exposures," *Radiation Protection Dosimetry* 83, no. 1–2 (1999): 9–14, https://doi.org/10.1093/oxfordjournals.rpd.a032670.

65   T. Lobos et al., "Power Distortion Issues in Wind Turbine Power Systems Under Transient States," *Turkish Journal of Electrical Engineering & Computer Sciences* 16 (2008): 229–38, https://journals.tubitak.gov.tr/cgi/viewcontent.cgi?article=3546&context=elektrik.

66   Hillman et al., "Relationship of Electric Power Quality to Milk Production," 500–14.

67   S. Wan et al., "Transmittance of Nonionizing Radiation in Human Tissues," *Photochemistry and Photobiology* 34, no. 6 (1981): 679–81, https://doi.org/10.1111/j.1751-1097.1981.tb09063.x.

68   S. Milham, "Historical Evidence That Electrification Caused the 20th Century Epidemic of 'Diseases of Civilization,'" *Medical Hypotheses* 74, no. 2 (2010): 337–45, https://doi.org/10.1016/j.mehy.2009.08.032.

69   R. Hajar, "Animal Testing and Medicine," *Heart Views* 12, no. 1 (2011): 42, https://doi.org/10.4103/1995-705X.81548.

70   C. F. Blackman et al., "Multiple Power-Density Windows and Their Possible Origin," *Bioelectromagnetics* 1989, 10(2): 115–28, https://doi.org/10.1002/bem.2250100202.

71   L. Zastko et al., "Intermittent ELF-MF Induce an Amplitude-Window Effect on Umbilical Cord Blood Lymphocytes," *International Journal of Molecular Sciences* 23 (2022): 14391.

72   S. S. Ahmadi et al., "Effect of Non-ionizing Electromagnetic Field on the Alteration of Ovarian Follicles in Rats," *Electronic Physician* 8, no. 3 (2016): 2168–74, https://doi.org/10.19082/2168.

73   S. I. Alekperov et al., "The Effect of Electromagnetic Fields of Extremely Low Frequency 30 Hz on Rat Ovaries," *Bulletin of Experimental*

Biology and Medicine 166 (2019): 704–7, https://doi.org/10.1007/s10517-019-04422-2.

74   D. O. Carpenter, "Extremely Low Frequency Electromagnetic Fields and Cancer: How Source of Funding Affects Results," *Environmental Research* 178 (2019): 108688, https://doi.org/10.1016/j.envres.2019.108688.

75   N. Wertheimer and E. Leeper, "Electrical Wiring Configurations and Childhood Cancer," *American Journal of Epidemiology* 109, no. 3 (1979): 273–84, https://doi.org/10.1093/oxfordjournals.aje.a112681.

76   N. Wertheimer and E. Leeper, "Adult Cancer Related to Electrical Wires Near the Home," *International Journal of Epidemiology* 11, no. 4 (1982): 345–55, https://doi.org/10.1093/ije/11.4.345.

77   Milham, "Historical Evidence That Electrification Caused the 20th Century Epidemic," 337–45.

78   M. Terzi et al., "The Role of Electromagnetic Fields in Neurological Disorders," *Journal of Chemical Neuroanatomy* 75, no. B (2016): 77–84, https://doi.org/10.1016/j.jchemneu.2016.04.003.

79   A. Karenberg and H. Forstl, "Dementia in the Greco-Roman World," *Journal of Neurological Sciences* 244, no. 1–2 (2006): 5–9, https://doi.org/10.1016/j.jns.2005.12.004.

80   V. C. Hachinski et al., "Multi-infarct Dementia: A Cause of Mental Deterioration in the Elderly," *The Lancet* 304, no. 7874 (1974): 207–9, https://doi.org/10.1016/S0140-6736(74)91496-2.

81   D. Perl, "Relationship of Aluminum to Alzheimer's Disease," *Environmental Health Perspectives* 63 (1985), https://doi.org/10.1289/ehp.8563149.

82   M. B. Suárez-Fernández et al., "Aluminum-Induced Degeneration of Astrocytes Occurs via Apoptosis and Results in Neuronal Death," *Brain Research* 835, no. 2 (1999): 125–36, https://doi.org/10.1016/S0006-8993(99)01536-X.

83   Y. Deng et al., "Effects of Aluminum and Extremely Low Frequency Electromagnetic Radiation on Oxidative Stress and Memory in Brain of Mice," *Biological Trace Element Research* 156 (2013): 243–52, https://doi.org/10.1007/s12011-013-9847-9.

84   N. H. Hansen et al., "EMF Exposure Assessment in the Finnish Garment Industry: Evaluation of Proposed EMF Exposure Metrics," *Bioelectromagnetics* 21, no. 1 (2000 Jan): 57–67, https://doi.org/10.1002/(sici)1521-186x(200001)21:1<57::aid-bem9>3.0.co;2-p.

85   C. Qiu et al., "Occupational Exposure to Electromagnetic Fields and Risk of Alzheimer's Disease," *Epidemiology* 15, no. 6 (2004): 687–94, https://doi.org/10.1097/01.ede.0000142147.49297.9d.

86   A. Seidler et al., "Occupational Exposure to Low Frequency Magnetic Fields and Dementia: A Case-Control Study," *Occupational and*

Environmental Medicine 64 (2007): 108–14, https://doi.org/10.1136/oem.2005.024190.

87   H. Jalilian et al., "Occupational Exposure to Extremely Low Frequency Magnetic Fields and Risk of Alzheimer Disease: A Systematic Review and Meta-Analysis," *Neurotoxicology* 69 (2018): 242–52, https://doi.org/10.1016/j.neuro.2017.12.005.

88   C. M. Calvin et al., "Association of Multimorbidity, Disease Clusters, and Modification by Genetic Factors with Risk of Dementia," *JAMA Network Open* 5, no. 9 (September 1, 2022): e2232124, https://doi.org/10.1001/jamanetworkopen.2022.32124.

89   J. M. Ortman et al., "An Aging Nation: The Older Population in the United States," United States Census Bureau, 2014: P25–1140.

90   P. T. Kotzbauer et al., "Lewy Body Pathology in Alzheimer's Disease," *Journal of Molecular Neuroscience* 17 (2001): 225–32, https://doi.org/10.1385/JMN:17:2:225.

91   K. Ubhi et al., "Multiple System Atrophy: A Clinical and Neuropathological Perspective," *Trends in Neuroscience* 34, no. 11 (2012): 581–90, https://doi.org/10.1016/j.tins.2011.08.003.

92   W. S. Kim et al., "Alpha-synuclein Biology in Lewy Body Diseases," *Alzheimer's Research & Therapy* 6, no. 5 (2014): 73, https://doi.org/10.1186/s13195-014-0073-2.

93   C. Pritchard et al., "Are Rises in Electro-magnetic Field in the Human Environment, Interacting with Multiple Environmental Pollutions, the Tipping Point for Increases in Neurological Deaths in the Western World?" *Medical Hypotheses* 127 (2019): 76–83, https://doi.org/10.1016/j.mehy.2019.03.018.

94   A. Huss et al., "Occupational Exposure to Extremely Low-Frequency Magnetic Fields and the Risk of ALS: A Systematic Review and Meta-Analysis," *Bioelectromagnetics.* 39, no. 2 (2018): 156–63, https://doi.org/10.1002/bem.22104.

95   H. Jalilian et al., "Amyotrophic Lateral Sclerosis, Occupational Exposure to Extremely Low Frequency Magnetic Fields and Electric Shocks: A Systematic Review and Meta-Analysis," *Reviews on Environmental Health* 36, no. 1 (2020): 129–42, https://doi.org/10.1515/reveh-2020-0041.

96   P. Jenner et al., "Oxidative Stress as a Cause of Nigral Cell Death in Parkinson's Disease and Incidental Lewy Body Disease," *Annals of Neurology* 32 (1992): S82–S87, https://doi.org/10.1002/ana.410320714.

97   B. R. De Miranda and J. T. Greenamyre, "Trichloroethylene, a Ubiquitous Environmental Contaminant in the Risk for Parkinson's Disease," *Environmental Science: Processes & Impacts* 22, no. 3 (2020): 543–54, https://doi.org/10.1039/c9em00578a.

98   A. Huss et al., "Extremely Low Frequency Magnetic Field Exposure and Parkinson Disease – a Systematic Review and Meta-Analysis of the Data," *International Journal of Environmental Research and Public Health* 12, no. 7 (2015): 7348–56, https://doi.org/10.3390/ijerph120707348.

99   F. Khaki-Khatibi et al., "Relationship Between the Use of Electronic Devices and Susceptibility to Multiple Sclerosis," *Cognitive Neurodynamics* 13 (2019): 287–92, https://doi.org/10.1007/s11571-019-09524-1.

100 A. Ahlbom et al., "A Pooled Analysis of Magnetic Fields and Childhood Leukaemia," *British Journal of Cancer* 83, no. 5 (September 2000): 692–98, https://doi.org/10.1054/bjoc.2000.1376.

101 S. Greenland et al., "A Pooled Analysis of Magnetic Fields, Wire Codes, and Childhood Leukemia, Childhood Leukemia-EMF Study Group," *Epidemiology* 11, no. 6 (November 2000): 624–34, https://doi.org/10.1097/00001648-200011000-00003.

102 D. O. Carpenter, "Extremely Low Frequency Electromagnetic Fields and Cancer: How Source of Funding Affects Results," *Environmental Research* 178 (2019): 108688, https://doi.org/10.1016/j.envres.2019.108688.

103 W. S. Burrup et al., "Low Frequency Electromagnetic Field Suppresses Macrophage Phagocytosis [abstract]," *Cancer Research* 89, no. 13 Suppl (2018): Abstract 5692.

104 C. Yao et al., "The Biological Effects of Electromagnetic Exposure on Immune Cells and Potential Mechanisms," *Electromagnetic Biology and Medicine* 41, no. 1 (2022): 108–17, https://doi.org/10.1080/15368378.2021.2001651.

105 P. R. Doyon and O. Johansson, "Electromagnetic Fields May Act via Calcineurin Inhibition to Suppress Immunity, Thereby Increasing Risk for Opportunistic Infection: Conceivable Mechanisms of Action," *Medical Hypotheses* 106 (2017): 71–87, https://doi.org/10.1016/j.mehy.2017.06.028.

106 P. Piszczek et al., "Immunity and Electromagnetic Fields," *Environmental Research* 200 (2021): 111505, https://doi.org/10.1016/j.envres.2021.111505.

107 B. J. Scherlag et al., "Magnetism and Cardiac Arrhythmias," *Cardiology in Review* 12, no. 2 (2004): 85–96, https://doi.org/10.1097/01.crd.0000094029.10223.2f.

108 J. Kornej et al., "Epidemiology of Atrial Fibrillation in the 21st Century: Novel Methods and New Insights," *Circulation Research* 127, no. 1 (2020): 4–20, https://doi.org/10.1161/CIRCRESAHA.120.316340.

109 N. Todorova et al., "Electromagnetic Field Modulates Aggregation Propensity of Amyloid Peptides," *The Journal of Chemical Physics* 152 (2020): 035104, https://doi.org/10.1063/1.5126367.

110 Todorova et al., "Electromagnetic Field Modulates Aggregation Propensity of Amyloid Peptides."

111 S. Milham and L. L. Morgan, "A New Electromagnetic Exposure Metric: High Frequency Voltage Transients Associated with Increased Cancer Incidence in Teachers in a California School," *American Journal of Industrial Medicine* 51, no. 8 (2008): 579–86, https://doi.org/10.1002/ajim.20598.

112 S. Milham, "Attention Deficit Hyperactivity Disorder and Dirty Electricity," *Journal of Developmental & Behavioral Pediatrics* 32, no. 8 (2011): 634, https://doi.org/10.1097/DBP.0b013e31822f8da7.

113 M. Havas and A. Olstad, "Power Quality Affects Teacher Wellbeing and Student Behavior in Three Minnesota Schools," *Science of the Total Environment* 402, no. 2–3 (2008): 157–62, https://doi.org/10.1016/j.scitotenv.2008.04.046.

114 M. Havas, "Dirty Electricity Elevates Blood Sugar Among Electrically Sensitive Diabetics and May Explain Brittle Diabetes," *Electromagnetic Biology and Medicine* 27, no. 2 (2008): 135–46, https://doi.org/10.1080/15368370802072075.

115 S. Milham, "Evidence That Dirty Electricity Is Causing the Worldwide Epidemics of Obesity and Diabetes," *Electromagnetic Biology and Medicine* 33, no. 1 (2014): 75–78, https://doi.org/10.3109/15368378.2013.783853.

116 "Building Biology Evaluation Guidelines," Institute of Building Biology + Sustainability IBN, accessed March 10, 2026, https://www.buildingbiology.com; "Building Biology Evaluation Guidelines for Sleeping Areas," *Baubiologie Maes*, accessed March 10, 2026, https://www.maes.de.

117 BioInitiative Working Group, *The Bioinitiative Report: A Rationale for a Biologically-based Public Exposure Standard for Electromagnetic Fields (ELF and RF) 2012*, ed. Cindy Sage and David O. Carpenter, updated 2014–2022, https://bioinitiative.org/.

118 ICNIRP, "International Commission on Non-Ionizing Radiation Protection (ICNIRP) Guidelines on Limits of Exposure to Static Magnetic Fields," *Health Physics* 96, no. 4 (2009): 504–14.

119 C.-F. Yang et al., "Mitigation of Magnetic Field Using Three-Phase Four-Wire Twisted Cables," *International Transactions on Electrical Energy Systems* 23 (2013): 13–23, https://doi.org/10.1002/etep.636/pdf.

120 K. Zhu et al., "Prostate Cancer in Relation to the Use of Electric Blanket or Heated Water Bed," *Epidemiology* 10, no. 1 (1999): 83–85, https://doi.org/10.1097/00001648-199901000-00016.

121 F. H. Krusen et al., "Microkymatotherapy: Preliminary Report of Experimental Studies of the Heating Effect of Microwaves (Radar) in

Living Tissues," *Proceedings of the Mayo Clinic* 22 (1940): 212, PMID: 20241418.

122 K. G. Wakim et al., "Therapeutic Possibilities of Microwaves; Experimental and Clinical Investigation," *Journal of the American Medical Association* 139 (1949): 992, https://doi.org/10.1001/jama.1949 .02900320019006.

123 J. C. King, "Therapeutic Fever Produced by Diathermy – Its Present Developments and Future Possibilities," *Radiology* 20, no. 6 (1933): 449–56, https://doi.org/10.1148/20.6.449.

124 A. Gosset et al., "Essai de Thérapeutique de 'Cancer Experimental' des Plantes," *Comtes Rendus de la Société de Biologie* 91 (1924): 626–28.

125 I. E. Dodueva et al., "Plant Tumors: A Hundred Years of Study," *Planta* 251, no. 4 (2020): 82, https://doi.org/10.1007/s00425-020-03375-5.

126 H. J. Cook et al., "Early Research on the Biological Effects of Microwave Radiation: 1940–1960," *Annals of Science* 37 (1980): 323–51, https://doi .org/10.1080/00033798000200271.

127 R. V. Christie and A. L. Loomis, "Relationship of Frequency to Physiological Effects of Ultra-high Frequency Currents," *Journal of Experimental Medicine* 49 (1929): 321.

128 R. C. Dorf, *The Electrical Engineering Handbook, Second Edition* (CRC Press, 1993), 1046.

129 L. E. Daily, "A Clinical Study of the Results of Exposure of Laboratory Personnel to Radar and High Frequency Radio," *U.S. Naval Medical Bulletin* 41 (1943): 1052–56.

130 Z. Glaser, *Bibliography of Reported Biological Phenomena ('Effects') and Clinical Manifestations Attributed to Microwave and Radio-Frequency Radiation, Second Printing* (Naval Medical Research Institute, 1972), https://zoryglaser.com/wp-content/uploads/2020/05/A-CLINICAL -STUDY-OF-THE-RESULTS-OF-EXPOSURE-OF-LABORATORY -PERSONNEL-TO-RADAR-AND-HIGH-FREQUENCY-RADIO -1.pdf.

131 Quoted in B.I. Lidman and C. Cohn, "Effect of Radar Emanations on the Hematopoietic System," *Air Surgeon's Bulletin* 2 (1945): 449.

132 W. W. Salisbury et al., "Exposure to Microwaves," *Electronics* 22 (1949): 66.

133 Unpublished Navy Department conference, April 29, 1953, Naval Medical Research Institute, Bethesda, Maryland, 5–8.

134 J. R. McLaughlin, "A Survey of Possible Health Hazards from Exposure to Microwave Radiation," *Hughes Aircraft Corp.* 1953, Culver City, Calif: 5–6.

135 H. P. Schwan and K. Li, "The Mechanism of Absorption of Ultrahigh Frequency Electromagnetic Energy in Tissues, as Related to the Problem

of Tolerance Dosage," *IRE Transactions on Medical Electronics ME-4* 1956, 49, https://doi.org/10.1109/IRET-ME.1956.5008561.

136 K. R. Foster, "In Memoriam: Herman P. Schwan [1915–2005]," *Biomedical Engineering Online* 4 (2005): 21, https://doi.org/10.1186/1475-925X-4-21.

137 C. I. Barron et al., "Physical Evaluation of Personnel Exposed to Microwave Emanations," *Journal of Aviation Medicine* 26 (1955): 442, https://doi.org/10.1109/IRET-ME.1956.5008560.

138 A. A. Teixeira-Pinto et al., "The Behavior of Unicellular Organisms in an Electromagnetic Field," *Experimental Cell Research* 20 (1960): 548–64, https://doi.org/10.1016/0014-4827(60)90123-3.

139 U.S. Air Force, "Electromagnetic Radiation Hazards," T.O. 31Z-10-4, 1966; rev. 1967, U.S. Senate Committee on Commerce, Hearing on S. 2067 (90th Cong., 1st sess., 1968): 407–8.

140 U.S. Senate Committee on Commerce, Hearing on S. 2067 (90th Cong., 1st sess., 1968), 407–8.

141 N. H. Steneck et al., "The Origins of U.S. Safety Standards for Microwave Radiation," *Science* 208 (1980): 1230–36, https://doi.org/10.1126/science.6990492.

142 Glaser, *Bibliography of Reported Biological Phenomena*, 1–103.

143 "About Us," Omdia, accessed March 10, 2026, https://omdia.tech.informa.com/about-us; "Exploding 5G Adoption Continues Around the World," 5G Americas, December 15, 2022, https://www.5gamericas.org/exploding-5g-adoption-continues-around-the-world/.

144 "Exploding 5G Adoption Continues Around the World," *Microwave Journal*, December 15, 2022, https://www.microwavejournal.com/articles/39428-exploding-5g-adoption-continues-around-the-world.

145 "Mobile and PSTN Communication Services: Competition or Complementarity?" *OECD Digital Economy Papers*, no. 13, January 1, 1995, https://doi.org/10.1787/237485605680.

146 "Forecast: Global 2G Mobile Subscribers 2008–2020," *Statista Research Department*, September 8, 2011, https://www.statista.com/statistics/226085/global-2g-mobile-subscriber-forecast.

147 R. R. Brown, "Assessment of Radiofrequency Radiation Intensity on 35 Main Streets Throughout Pennsylvania, USA During the Fall of 2021," *Asian Journal of Multidisciplinary Research & Review* 1, no. 4 (2022): 8–20, https://www.ajmrr.com/wp-content/uploads/2022/12/B140820.pdf.

148 E. Arias et al., "Provisional Life Expectancy Estimates for 2021," *Vital Statistics Rapid Release*, no. 23 (2022).

149 B. W. Ward and J. S. Schiller, "Prevalence of Multiple Chronic Conditions among US Adults: Estimates from the National Health Interview Survey,

2010," *Preventing Chronic Disease* 10 (2013): E65, https://doi.org/10.5888/pcd10.120203.

150 P. Boersma et al., "Prevalence of Multiple Chronic Conditions among US Adults, 2018," *Preventing Chronic Disease* 17 (2020): E106, https://doi.org/10.5888/pcd17.200130.

151 D. M. Qato et al., "Prescription Medication Use among Children and Adolescents in the United States," *Pediatrics* 142, no. 3 (2018): e20181042, https://doi.org/10.1542/peds.2018-1042.

152 O. Hallberg, "Cancer Incidence vs. FM Radio Transmitter Density," *Electromagnetic Biology and Medicine* 35, no. 4 (2016), https://doi.org/10.3109/15368378.2016.1138122.

153 A. Huss et al., "Source of Funding and Results of Studies of Health Effects of Mobile Phone Use: Systematic Review of Experimental Studies," *Environmental Health Perspectives* 115, no. 1 (2007): 1–4, https://doi.org/10.1289/ehp.9149.

154 D. J. Panagopoulos, "Comparing DNA Damage Induced by Mobile Telephony and Other Types of Man-made Electromagnetic Fields," *Mutation Research* 781 (2019): 53–62, https://doi.org/10.1016/j.mrrev.2019.03.003.

155 J. Sliny, "Demodulation in Tissue, the Relevant Parameters and the Implications for Limiting Exposure," *Health Physics* 92, no. 6 (2007): 604–8, https://doi.org/10.1097/01.HP.0000244086.36815.7c.

156 C. Palombini, "Electromagnetic Precursors in Complex Layered Media," Ph.D. Dissertation, Electrical Engineering, University of Vermont, Graduate College, 2012, https://scholarworks.uvm.edu/graddis/175/.

157 K. R. Foster et al., "Thermal Response of Human Skin to Microwave Energy: A Critical Review," *Health Physics* 111, no. 6 (2016): 528–41, https://doi.org/10.1097/HP.0000000000000571.

158 S. I. Alekseev et al., "Millimeter Wave Dosimetry of Human Skin," *Bioelectromagnetics* 29, no. 1 (2008): 65–70, https://doi.org/10.1002/bem.20363.

159 A. Y. Owda et al., "The Reflectance of Human Skin in the Millimeter-Wave Band," *Sensors (Basel)* 20, no. 5 (2020): 1480, https://doi.org/10.3390/s20051480.

160 N. Betzalel et al., "The Human Skin as a Sub-THz Receiver: Does 5G Pose a Danger to It or Not?" *Environmental Research* 163 (2018): 208–16, https://doi.org/10.1016/j.envres.2018.01.032.

161 G. Sacco et al., "Impact of Textile on Electromagnetic Power and Heating in Near-Surface Tissues at 26 GHz and 60 GHz," *IEEE Journal of Electromagnetics, RF and Microwaves in Medicine and Biology* 5, no. 3 (2021): 262–68, https://doi.org/10.1109/JERM.2020.3042390.

162 K. Kim et al., "5G Electromagnetic Radiation Attenuates Skin Melanogenesis in Vitro by Suppressing ROS Generation," *Antioxidants* 11, no. 8 (2022): 1449, https://doi.org/10.3390/antiox11081449.

163 J. H. Han et al., "Mass Screening on Comorbidities of Vitiligo: A Nationwide Population-Based Study," *Proceedings: The 73rd Korean Dermatological Society Meeting* 72, no. 1 (2020): 441–42, https://kiss.kstudy.com/thesis/thesis-view.asp?key=3845245.

164 H. H. Balik et al., "Effets Oculaires en Relation avec l'Utilisation d'un Téléphone Mobile Cellulaire," *Pathologie Biologie* 53, no. 2 (2005): 88–91, https://doi.org/10.1016/j.patbio.2004.03.012.

165 H. A. Kues and J. C. Monahan, "Microwave-Induced Changes to the Primate Eye," *Johns Hopkins APL Technical Digest* 13, no. 1 (1992): 244–55.

166 E. Bormusov et al., "Non-thermal Electromagnetic Radiation Damage to Lens Epithelium," *Open Ophthalmology Journal* 2 (2008): 102–6, https://doi.org/10.2174/1874364100802010102.

167 S. Lixia et al., "Effects of 1.8 GHz Radiofrequency Field on DNA Damage and Expression of Heat Shock Protein 70 in Human Lens Epithelial Cells," *Molecular Mechanisms of Mutagenesis* 602, no. 1–2 (2006): 135–42, https://doi.org/10.1016/j.mrfmmm.2006.08.010.

168 Kues and Monahan, "Microwave-Induced Changes to the Primate Eye," 244–55.

169 R. M. Lipman et al., "Cataracts Induced by Microwave and Ionizing Radiation," *Survey of Ophthalmology* 33, no. 3 (1988): 200–10, https://doi.org/10.1016/0039-6257(88)90088-4.

170 K. R. Foster et al., "Nonuniform Exposure to the Cornea from Millimeter Waves," *Health Physics* 120, no. 5 (2021): 525–31, https://doi.org/10.1097/HP.0000000000001376.

171 J. Jean et al., "Laser-Induced Corneal Injury: Validation of a Computer Model to Predict Thresholds," *Biomedical Optics Express* 12, no. 1 (2020): 336–53, https://doi.org/10.1364/BOE.412102.

172 M. A. Esmekaya et al., "Pulse Modulated 900 MHz Radiation Induces Hypothyroidism and Apoptosis in Thyroid Cells: A Light, Electron Microscopy and Immunohistochemical Study," *International Journal of Radiation Biology* 86 (2010): 1106–16, https://doi.org/10.3109/09553002.2010.502960.

173 N. M. Baby et al., "The Effect of Electromagnetic Radiation due to Mobile Phone Use on Thyroid Function in Medical Students Studying in a Medical College in South India," *Indian Journal of Endocrinology and Metabolism* 21, no. 6 (2017): 797–802, https://doi.org/10.4103/ijem.IJEM_12_17.

174 J. F. Asi et al., "The Possible Global Hazard of Cell Phone Radiation on Thyroid Cells and Hormones: A Systematic Review of Evidences," *Environmental Science and Pollution Research* 26 (2019): 18017–31, https://doi.org/10.1007/s11356-019-05096-z.

175 T. Alkayyali et al., "An Exploration of the Effects of Radiofrequency Radiation Emitted by Mobile Phones and Extremely Low Frequency Radiation on Thyroid Hormones and Thyroid Gland Histopathology," *Cureus* 13, no. 8 (2021): e17329, https://doi.org/10.7759/cureus.17329.

176 Alkayyali et al., "Exploration of the Effects of Radiofrequency Radiation."

177 J. Feng et al., "Epidemiological Characteristics of Infertility, 1990–2021, and 15-Year Forecasts: An Analysis Based on the Global Burden of Disease Study 2021," *Reproductive Health* 22 (2025): 26, https://doi.org/10.1186/s12978-025-01966-7.

178 R. Sciori et al., "Effects of Mobile Phone Radiofrequency Radiation on Sperm Quality," *Zygote* 30, no. 2 (2022): 159–68, https://doi.org/10.1017/S096719942100037X.

179 P. Negi and R. Singh, "Association Between Reproductive Health and Nonionizing Radiation Exposure," *Electromagnetic Biology and Medicine* 40, no. 1 (2021): 92–102, https://doi.org/10.1080/15368378.2021.1874973.

180 F. Scrinicariello and M. C. Buser, "Serum Testosterone Concentrations and Urinary Bisphenol A, Benzophenone-3, Triclosan, and Paraben Levels in Male and Female Children and Adolescents: NHANES 2011–2012," *Environmental Health Perspectives* 124, no. 12 (2016): CID, https://doi.org/10.1289/EHP150.

181 F. H. F. Jaffar et al., "Adverse Effects of Wi-Fi Radiation on Male Reproductive System: A Systematic Review," *The Tohoku Journal of Experimental Medicine* 248, no. 3 (2019): 169–79, https://doi.org/10.1620/tjem.248.169.

182 C. E. Okechukwu, "Does the Use of Mobile Phone Affect Male Fertility? A Mini-Review," *Journal of Human Reproductive Sciences* 13, no. 3 (2020): 174–83, https://doi.org/10.4103/jhrs.JHRS_126_19.

183 Jaffar et al., "Adverse Effects of Wi-Fi Radiation on Male Reproductive System," 169–79.

184 J. M. Gray et al., "State of the Evidence 2017: An Update on the Connection Between Breast Cancer and the Environment," *Environmental Health* 16 (2017): 94, https://doi.org/10.1186/s12940-017-0287-4.

185 J. G. West et al., "Multifocal Breast Cancer in Young Women with Prolonged Contact Between Their Breasts and Their Cellular Phones," *Case Reports in Medicine* ID 354682 (2013), https://doi.org/10.1155/2013/354682.

186 D. R. Nithiya et al., "Mobile Keeping Behavior Among Working Women," *Medico-Legal Update* 20, no. 4 (2020): 179, https://doi.org/10.37506/mlu.v20i4.1787.

187 Y.-W. Shih et al., "The Association Between Smartphone Use and Breast Cancer Risk Among Taiwanese Women: A Case-Control Study," *Cancer Management and Research* 12 (2020): 10799–10807, https://doi.org/10.2147/CMAR.S267415.

188 B. Selmaoui and Y. Touitou, "Association Between Mobile Phone Radiation Exposure and the Secretion of Melatonin and Cortisol, Two Markers of the Circadian System: A Review," *Bioelectromagnetics* 42, no. 1 (2020): 5–17, https://doi.org/10.1002/bem.22310.

189 L. Pistioli et al., "The Intricate Relationship Between Melatonin and Breast Cancer: A Short Review," *Chirugia* 116, no. 2 (2021): S24–S34, https://doi.org/10.21614/chirurgia.116.2.

190 C. Yu and R.-Y. Peng, "Biological Effects and Mechanisms of Shortwave Radiation: A Review," *Military Medical Research* 4 (2017): 24, https://doi.org/10.1186/s40779-017-0133-6.

191 Selmaoui and Touitou, "Association Between Mobile Phone Radiation Exposure and the Secretion of Melatonin and Cortisol," 5–17.

192 A. Shehu et al., "Exposure to Mobile Phone Electromagnetic Field Radiation, Ringtone and Vibration Affects Anxiety-like Behaviour and Oxidative Stress Biomarkers in Albino Wistar Rats," *Metabolic Brain Disease* 31 (2016): 355–62, https://doi.org/10.1007/s11011-015-9758-x.

193 R. C. Vera and I. Muñoz, "The Influence of Electromagnetic Fields on the Behavior of Mice," *IntechOpen* 2021, https://doi.org/10.5772/intechopen.93320.

194 T. W. Miller, "School-Related Violence: Definition, Scope, and Prevention Goals," in *School Violence and Primary Prevention* (Springer, Cham, 2023), 3–18, https://doi.org/10.1007/978-3-031-13134-9_1.

195 P. Slottje et al., "Electromagnetic Hypersensitivity (EHS) in Occupational and Primary Health Care: A nation-wide Survey Among General Practitioners, Occupational Physicians and Hygienists in the Netherlands," *International Journal of Hygiene and Environmental Health* 220, no. 2 Pt B (2017): 395–400, https://doi.org/10.1016/j.ijheh.2016.11.013.

196 A. Barth et al., "No Effects of Short-Term Exposure to Mobile Phone Electromagnetic Fields on Human Cognitive Performance: A Meta-Analysis," *Bioelectromagnetics* 33, no. 2 (2012): 159–65, https://doi.org/10.1002/bem.20697.

197 J. Tang et al., "Exposure to 900 MHz Electromagnetic Fields Activates the mkp-1/ERK Pathway and Causes Blood-Brain Barrier Damage and Cognitive Impairment in Rats," *Brain Research* 1601 (2015): 92–101, https://doi.org/10.1016/j.brainres.2015.01.019.

198 B. Pophof et al., "The Effect of Exposure to Radiofrequency Electromagnetic Fields on Cognitive Performance in Human Experimental Studies: A Protocol for a Systematic Review," *Environ International* 157 (2021): 106783, https://doi.org/10.1016/j.envint.2021.106783.

199 L. Hardell and M. Nilsson, "Case Report: The Microwave Syndrome After Installation of 5G Emphasizes the Need for Protection from Radiofrequency Radiation," *Annals of Case Reports* 8 (2023): 1112, https://doi.org/10.29011/2574-7754.101112.

200 J. Wang et al., "Mobile Phone Use and the Risk of Headache: A Systematic Review and Meta-Analysis of Cross-Sectional Studies," *Scientific Reports* 7 (2017): 12595, https://doi.org/10.1038/s41598-017-12802-9.

201 National Toxicology Program (NTP), *Toxicology and Carcinogenesis Studies of GSM- and CDMA-Modulated Cell Phone Radiofrequency Radiation at 900 MHz in Sprague Dawley Rats (Whole Body Exposure)*, Technical Report Series No. 595 (The National Toxicology Program, 2018).

202 National Toxicology Program (NTP), *Toxicology and Carcinogenesis Studies of GSM- and CDMA-Modulated Cell Phone Radiofrequency Radiation at 1,900 MHz in B6C3F1/N Mice (Whole Body Exposure)*, Technical Report Series No. 596 (The National Toxicology Program, 2018).

203 L. Falconi et al., "Report of Final Results Regarding Brain and Heart Tumors in Sprague-Dawley Rats Exposed from Prenatal Life Until Natural Death to Mobile Phone Radiofrequency Field Representative of a 1.8 GHz GSM Base Station Environmental Emission," *Environmental Research* 165 (2018): 496–503, https://doi.org/10.1016/j.envres.2018.01.037.

204 P. J. Villeneuve et al., "Cell Phone Use and the Fisk of Glioma: Are Case-Control Study Findings Consistent with Canadian Time Trends in Cancer Incidence?" *Environmental Research* 22 (2021): 111283, https://doi.org/10.1016/j.envres.2021.111283.

205 A. Bortkiewicz et al., "Mobile Phone Use and Risk for Intracranial Tumors and Salivary Gland Tumors—A Meta-Analysis," *International Journal of Occupational Medicine and Environmental Health* 30, no. 1 (2017): 27–43, https://doi.org/10.13075/ijomeh.1896.00802.

206 M. Röösli et al., "Brain and Salivary Gland Tumors and Mobile Phone Use: Evaluating the Evidence from Various Epidemiological Study Designs," *Annual Review of Public Health* 40 (2019): 221–38, https://doi.org/10.1146/annurev-publhealth-040218-044037.

207 J. Schuz et al., "Cellular Phones, Cordless Phones, and the Risks of Glioma and Meningioma (Interphone Study Group, Germany)," *American Journal of Epidemiology* 163 (2006): 512–20, https://doi.org/10.1093/aje/kwj068.

208 S. Lönn et al., "Mobile Phone Use and the Risk of Acoustic Neuroma,"
     Epidemiology 15 (2004): 653–59, https://doi.org/10.1097/01.ede
     .0000142519.00772.bf.

209 L. Hardell et al., "Case-Control Study of the Association Between
     Malignant Brain Tumours Diagnosed Between 2007 and 2009 and Mobile
     and Cordless Phone Use," *International Journal of Oncology* 43 (2013):
     1833–45, https://doi.org/10.3892/ijo.2013.2111.

210 Röösli et al., "Brain and Salivary Gland Tumors and Mobile Phone Use,"
     221–38.

211 Y.-J. Choi et al., "Cellular Phone Use and Risk of Tumors: Systematic
     Review and Meta-Analysis," *International Journal of Environmental
     Research and Public Health* 17, no. 21 (2020): 8079, https://doi.org/
     10.3390/ijerph17218079.

212 K. Karipidis et al., "The Effect of Exposure to Radiofrequency Fields on
     Cancer Risk in the General and Working Population: A Systematic Review
     of Human Observation Studies – Part 1: Most Researched Outcomes,"
     Environment International 191 (2024): 108983, https://doi.org/10.1016/
     j.envint.2024.108983.

213 J. W. Frank et al., "The Systematic Review on RF-EMF Exposure and
     Cancer by Karipidis et al. (2024) Has Serious Flaws That Undermine
     the Validity of the Study's Conclusions," *Environment International* 195
     (2025): 109200, https://doi.org/10.1016/j.envint.2024.109200.

214 B. Ekici et al., "The Effects of the Duration of Mobile Phone Use on Heart
     Rate Variability Parameters in Healthy Subjects," *The Anatolian Journal
     of Cardiology* 16, no. 11 (2016): 833–38, https://doi.org/10.14744/
     AnatolJCardiol.2016.6717.

215 J. Misek et al., "Heart Rate Variability Affected by Radiofrequency
     Electromagnetic Field in Adolescent Students," *Bioelectromagnetics* 39, no.
     4 (2018): 277–88, https://doi.org/10.1002/bem.22115.

216 J. Wallace et al., "Heart Rate Variability in Healthy Young Adults
     Exposed to Global System for Mobile Communication (GSM) 900-MHz
     Radiofrequency Signal from Mobile Phones," *Environmental Research* 191
     (2020): 110097, https://doi.org/10.1016/j.envres.2020.110097.

217 S. F. Cleary, "Biological Effects and Health Implications of Microwave
     Radiation–Symposium Proceedings, Richmond Virginia, September
     17–19, 1969," Department of Biophysics Virginia Commonwealth
     University, US Department of Health, Education, and Welfare.

218 Glaser, *Bibliography of Reported Biological Phenomena*, 106.

219 Z. R. Glaser et al., *Bibliography of Reported Biological Phenomena (Effects)
     and Clinical Manifestations Attributed to Microwave and Radio-frequency
     Radiation: Compilation and Integration of report and Seven Supplements,
     Revised Edition* (Naval Medical Research Institute, 1976), 1–178.

220 P. Bandara and S. Weller, "Cardiovascular Disease: Time to Identify Emerging Environmental Risk Factors," *European Journal of Preventive Cardiology* 24 (2017): 1819–23, https://doi.org/10.1177/2047487317734898.

221 J. R. Jauchem, "Exposure to Extremely Low Frequency Electromagnetic fields and radiofrequency Radiation: Cardiovascular Effects in Humans," *International Archives of Occupational and Environmental Health* 70 (1997): 9–21, https://doi.org/10.1007/s004200050181.

222 M. Havas et al., "Provocation Study Using Heart Rate Variability Shows Microwave Radiation from 2.4 GHz Cordless Phone Affects Autonomic Nervous System," in *Non-Thermal Effects and Mechanisms of Interaction Between Electromagnetic Fields and Living Matter: An ICEMS Monograph*, ed. Livio Giuliani and Morando Soffritti (2010), 273–300, https://bemri .org/publications/dect/341-provocation-study-using-heart-rate-variability -shows-microwave-radiation-from-2-4ghz-cordless-phone/file.

223 L. Saili et al., "Effects of Acute Exposure to WIFI Signals (2.45GHz) on Heart Variability and Blood Pressure in Albino Rabbits," *Environmental Toxicology and Pharmacology* 40, no. 2 (2015): 600–605, https://doi.org/10.1016/j.etap.2015.08.015.

224 D. Parizek et al., "Electromagnetic Fields–Do They Pose a Cardiovascular Risk?" *Physiological Research* 72, no. 2 (2023): 199–208, https://doi.org/10.33549/physiolres.934938.

225 D. R. Black and L. N. Heynick, "Radiofrequency Effects on Blood Cells, Cardiac, Endocrine, and Immunological Functions," *Bioelectromagnetics* 6 (2003): S187–S195, https://doi.org/10.1002/bem.10166.

226 Q. Fang et al., "An Investigation on the Effect of Extremely Low Frequency Pulsed Electromagnetic Fields on Human Electrocardiograms (ECGs)," *International Journal of Environmental Research and Public Health* 13, no. 11 (2016): 1171, https://doi.org/10.3390/ijerph13111171.

227 A. Alhusseiny et al., "Electromagnetic Energy Radiated from Mobile Phone Alters Electrocardiographic Records of Patients with Ischemic Heart Disease," *Annals of Medical & Health Sciences Research Journal* 2, no. 2 (2012): 146–51, https://doi.org/10.4103/2141-9248.105662.

228 Ekici et al., "Effects of the Duration of Mobile Phone Use on Heart Rate Variability," 833–38.

229 C. A. Morillo et al., "Atrial Fibrillation: The Current Epidemic," *Journal of Geriatric Cardiology* 14, no. 3 (2017): 195–203, https://doi.org/10.11909/j.issn.1671-5411.2017.03.011.

230 F. Nadeem et al., "Interference by Modern Smartphones and Accessories with Cardiac Pacemakers and Defibrillators," *Current Cardiology Reports* 24 (2022): 347–53, https://doi.org/10.1007/s11886-022-01653-0.

231 G.-B. Ha et al., "Safety Evaluation of Smart Scales, Smart Watches, and Smart Rings with Bioimpedance Technology Shows Evidence of Potential Interference in Cardiac Implantable Electronic Devices," *Heart Rhythm* 20, no. 4 (2023): 561–71, https://doi.org/10.1016/j.hrthm.2022.11.026.

232 C. Lennerz et al., "Modern Security Screening and Electromagnetic Interference with Cardiac Implantable Electronic Devices," *Journal of the American College of Cardiology* 75, no. 10 (2020): 1238–39, https://doi.org/10.1016/j.jacc.2020.01.012.

233 X. Castellon and V. Bogdanova, "Chronic Inflammatory Diseases and Endothelial Dysfunction," *Aging and Disease* 7, no. 1 (2016): 81–89, https://doi.org/10.14336/AD.2015.0803.

234 F. Xu et al., "Calreticulin Attenuated Microwave Radiation-Induced Human Microvascular Endothelial Cell Injury Through Promoting Actin Acetylation and Polymerization," *Cell Stress Chaperones* 22, no. 1 (2017): 89–97, https://doi.org/10.1007/s12192-016-0745-x.

235 Y. M. Moustafa et al., "Effects of Acute Exposure to the Radiofrequency Fields of Cellular Phones on Plasma Lipid Peroxide and Anti-oxidase Activities in Human Erythrocytes," *Journal of Pharmaceutical and Biomedical Analysis* 26 (2001): 605–8, https://doi.org/10.1016/S0731-7085(01)00492-7.

236 Zothansiama et al., "Impact of Radiofrequency Radiation on DNA Damage and Antioxidants in Peripheral Blood Lymphocytes of Humans Residing in the Vicinity of Mobile Phone Base Stations," *Electromagnetic Biology and Medicine* 2017, 36 (3): 295–305, https://doi.org/10.1080/15368378.2017.1350584.

237 Y. G. Grigoriev, "Evidence for Effects on the Immune System, Immune System and EMF RF," Bioinitiative Report, Section 8, 1–24, https://www.bioinitiative.org/wp-content/uploads/pdfs/sec08_2012_Evidence_%20Effects_%20Immune_System.pdf.

238 K. S. Nageswari et al., "Effect of Chronic Microwave Radiation on T Cell-Mediated Immunity in the Rabbit," *International Journal of Biometeorology* 35 (1991): 92–97, https://doi.org/10.1007/BF01087483.

239 D. Adang et al., "Results of a Long-Term Low-Level Microwave Exposure of Rats," *IEEE Transactions on Microwave Theory and Techniques* 57, no. 10 (2009): 2488–97, https://doi.org/10.1109/TMTT.2009.2029667.

240 Y. G. Grigoriev et al., "Autoimmune Processes After Long-Term Low-Level Exposure to Electromagnetic Fields (Experimental Results). Part I. Mobile Communications and Changes in Electromagnetic Conditions for the Population. Need for Additional Substantiation of Existing Hygienic Standards," *Biophysics* 55, no. 6 (2010): 1041–45, https://doi.org/10.1134/S0006350910060278.

241 O. Johansson, "Disturbance of the Immune System by Electromagnetic Fields – A Potentially Underlying Cause for Cellular Damage and Tissue Repair Reduction Which Could Lead to Disease and Impairment," *Pathophysiology* 16, no. 2–3 (2009): 157–77, https://doi.org/10.1016/j.pathophys.2009.03.004.

242 S. Szmigielski, "Reaction of the Immune System to Low-Level RF/MW Exposures," *Science of the Total Environment* 454–55 (2013): 393–400, https://doi.org/10.1016/j.scitotenv.2013.03.034.

243 S. Das et al., "Meta-Analysis of EMF-Induced Pollution by COVID-19 in Virtual Teaching and Learning with an Artificial Intelligence Perspective," *International Journal of Web-based Learning and Teaching Technologies* 17, no. 4 (2022): 1–20, https://doi.org/10.4018/IJWLTT.285566.

244 B. Rubik and R. Brown, "Evidence for a Connection Between Coronavirus Disease-19 and Exposure to Radiofrequency Radiation from Wireless Communications Including 5G," *Journal of Clinical and Translational Research* 7, no. 5 (September 29, 2021): 666–81, PMID: 34778597.

245 Panagopoulos, "Comparing DNA Damage Induced by Mobile Telephony," 53–62.

246 M. Peleg et al., "On Radar and Radio Exposure and Cancer in the Military Setting," *Environmental Research* 216, no. 2 (2023): 114610, https://doi.org/10.1016/j.envres.2022.114610.

247 Havas, "Radiation from Wireless Technology," 75–84.

248 C. Wagner et al., "Aggregation of Red Blood Cells: From Rouleaux to Clot Formation," *Comptes Rendus Physique* 14, no. 6 (2013): 459–69, https://doi.org/10.1016/j.crhy.2013.04.004.

249 B. Panyarachun et al., "The Decrease of Rouleaux Formation of Red Blood Cells in Healthy Human by Water-Soluble Chlorophyll as Revealed by Scanning Electron Microscopy," *Thai Pharmaceutical and Health Science Journal* 4, no. 4 (2009): 450–55, https://ejournals.swu.ac.th/index.php/pharm/article/view/2775.

250 S. Franzellitti et al., "Transient DNA Damage Induced by High-Frequency Electromagnetic Fields (GSM 1.8 GHz) in the Human Trophoblast HTR-8/SVneo Cell Line Evaluated with the Alkaline Comet Assay," *Mutation Research* 683, no. 1–2 (2010): 35–42, https://doi.org/10.1016/j.mrfmmm.2009.10.004.

251 S. Y. Perov et al., "Status of the Neuroendocrine System in Animals Chronically Exposed to Electromagnetic Fields of 5G Mobile Network Base Stations," *Bulletin of Experimental Biology and Medicine* 174 (2022): 277–79, https://doi.org/10.1007/s10517-023-05689-2.

252 J. A. Sheng et al., "The Hypothalamic-Pituitary-Adrenal Axis: Development, Programming Actions of Hormones, and Maternal-Fetal Interactions," *Behavioral Neuroscience* 14 (2020), https://doi.org/10.3389/fnbeh.2020.601939.

253 U. Bergqvist and E. Vogel, eds., *Possible Health Implications of Subjective Symptoms and Electromagnetic Fields: A Report by a European Group of Experts for the European Commission, DG V* (European Commission DG V. National Institute for Working Life, 1997), https://gupea.ub.gu.se /bitstream/handle/2077/4156/ah1997_19.pdf?sequence=1.

254 L. Hillert et al., "Cognitive Behavioral Therapy for Patients with Electric Sensitivity – a Multidisciplinary Approach in a Controlled Study," *Psychotherapy and Psychosomatics* 67 (1998): 302–10, https://doi.org /10.1159/000012295.

255 S. Eltiti et al., "Symptom Presentation in Idiopathic Environmental Intolerance with Attribution to Electromagnetic Fields: Evidence for a Nocebo Effect Based on Data Re-Analyzed from Two Previous Provocation Studies," *Frontiers in Psychology* 9 (2018): 1563, https://doi.org/10.3389 /fpsyg.2018.01563.

256 A.-K. Brascher et al., "Are Media Reports Able to Cause Somatic Symptoms Attributed to WiFi Radiation? An Experimental Test of the Negative Expectation Hypothesis," *Environmental Research* 156 (2017): 265–71, https://doi.org/10.1016/j.envres.2017.03.040.

257 Johansson et al., "Skin Changes in Patients Claiming to Suffer from 'Screen Dermatitis,'" 234–38.

258 D. Belpomme and P. Irigaray, "Electrohypersensitivity as a Newly Identified and Characterized Neurologic Pathological Disorder: How to Diagnose, Treat, and Prevent It," *International Journal of Molecular Science* 21, no. 6 (2020): 1915, https://doi.org/10.3390/ijms21061915.

259 P. Irigaray et al., "Oxidative Stress in Electrohypersensitivity Self-reporting Patients: Results of a Prospective in Vivo Investigation with Comprehensive Molecular Analysis," *International Journal of Molecular Medicine* (published online July 12, 2018): 1885–98, https://doi.org /10.3892/ijmm.2018.3774.

260 K. H. Mild et al., eds., *Electromagnetic Hypersensitivity: Proceedings of the International Workshop on Electromagnetic Field Hypersensitivity, Prague, Czech Republic, October 25–27, 2004* (World Health Organization, 2004), https://iris.who.int/handle/10665/43435.

261 "Report on Multiple Chemical Sensitivies (MCS) Workshop, Berlin Germany, 21–23 February 1996," International Programme on Chemical Safety/World Health Organization (IPCS/WHO), https://iris.who.int /server/api/core/bitstreams/94d38463-57da-444d-ad87-84090a779157 /content.

262 L. Hillert et al., "Prevalence of Self-reported Hypersensitivity to Electric or Magnetic Fields in a Population-Based Questionnaire Survey," *Scandinavian Journal of Work, Environment & Health* 28, no. 1 (2002): 33–41, https://doi.org/10.5271/sjweh.644.

263 J. Schröttner and N. Leitgeb, "Sensitivity to Electricity—Temporal
Changes in Austria," *BMC Public Health* 8 (2008): 310, https://doi.org
/10.1186/1471-2458-8-310.

264 P. Levallois et al., "Study of Self-Reported Hypersensitivity to
Electromagnetic Fields in California," *Environmental Health Perspectives*
110, no. Suppl. 4 (2002): 619–23.

265 M. Bevington, "The Prevalence of People with Restricted Access to Work
in Man-made Electromagnetic Environments," *Journal of Environmental
Health Sciences* 5, no. 1 (2019): 1–12, https://doi.org/10.15436/2378
-6841.19.2402.

266 I. van Moorselaar et al., "Effects of Personalised Exposure on Self-rated
Electromagnetic Hypersensitivity and Sensibility: A Double-Blind
Randomised Controlled Trial," *Environment International* 99 (2017):
255–62, https://doi.org/10.1016/j.envint.2016.11.031.

267 R. Wever, "The Effects of Electric Fields on Circadian Rhythmicity in
Men," *Life Sciences in Space Research* 8 (1970): 177–87, PMID: 11826883.

268 P. Héroux et al., "Cell Phone Radiation Exposure Limits and Engineering
Solutions," *International Journal of Environmental Research and Public
Health* 20, no. 7 (2023): 5398, https://doi.org/10.3390/ijerph20075398.

269 D. J. Panagopoulos and G. P. Chrousos, "Shielding Methods and Products
Against Man-made Electromagnetic Fields: Protection Versus Risk,"
Science of the Total Environment 667, no. 1 (2019): 255–62, https://doi
.org/10.1016/j.scitotenv.2019.02.344.

270 S. C. Cuthbert and G. J. Goodheart, "On the Reliability and Validity of
Manual Muscle Testing: A Literature Review," *Chiropractic & Osteopathy*
15 (2007): 4, https://doi.org/10.1186/1746-1340-15-4.

271 K. V. Kavokin, "Compass in the Ear: Can Animals Sense Magnetic Fields
with Hair Cells?" *The European Physical Journal Special Topics* 232 (2023):
261–68, https://doi.org/10.1140/epjs/s11734-022-00654-y.

272 J. L. Oschman et al., "The Effects of Grounding (Earthing) on
Inflammation, the Immune Response, Wound Healing, and Prevention
and Treatment of Chronic Inflammatory and Autoimmune Diseases,"
Journal of Inflammation Research 8 (2015): 83–96, https://doi.org
/10.2147/JIR.S69656.

273 G. Chevalier and S. T. Sinatra, "Emotional Stress, Heart Rate Variability,
Grounding, and Improved Autonomic Tone: Clinical Applications,"
Integrative Medicine 10, no. 3 (2011): 16–21, http://imjournal.com
/pdfarticles/imcj10_3_p16_24chevalier.pdf.

274 P. B. Lee et al., "Efficacy of Pulsed Electromagnetic Therapy for Chronic
Lower Back Pain: A Randomized, Double-Blind, Placebo-Controlled
Study," *The Journal of International Medical Research* 34, no. 2 (2006):
160–67, https://doi.org/10.1177/147323000603400205.

275 H. Wuschech et al., "Effects of PEMF on Patients with Osteoarthritis: Results of a Prospective, Placebo-Controlled, Double-Blind Study," *Bioelectromagnetics* 36, no. 8 (2015): 576–85, https://doi.org/10.1002/bem.21942.

276 D. E. Bragin et al., "Pulsed Electromagnetic Field (PEMF) Mitigates High Intracranial Pressure (ICP) Induced Microvascular Shunting (MVS) in Rats," in *Brain Edema XVI* (Springer International Publishing, 2018), 93–95, https://doi.org/10.1007/978-3-319-65798-1_20.

277 N. Biermann et al., "The Influence of Pulsed Electromagnetic Field Therapy on Lymphatic Flow During Supermicrosurgery," *Lymphatic Research and Biology* 18, no. 6 (2020): 549–54, https://doi.org/10.1089/lrb.2019.0094.

278 C. Marquez-Chin and M. R. Popovic, "Functional Electrical Stimulation Therapy for Restoration of Motor Function After Spinal Cord Injury and Stroke: A Review," *BioMedical Engineering OnLine* 19 (2020): 34, https://doi.org/10.1186/s12938-020-00773-4.

279 Y. Luo et al., "Deep Brain Stimulation for Alzheimer's Disease: Stimulation Parameters and Potential Mechanisms of Action," *Frontiers in Aging Neuroscience* 13 (2021), https://doi.org/10.3389/fnagi.2021.619543.

280 R. Sandyk, "Long Term Beneficial Effects of Weak Electromagnetic Fields in multiple Sclerosis," *International Journal of Neuroscience* 83 (1995): 45–57, https://doi.org/10.3109/00207459508986324.

281 J. P. Lefaucheur et al., "Evidence-Based Guidelines on the Therapeutic Use of Repetitive Transcranial Magnetic Stimulation (rTMS)," *Clinical Neurophysiology* 125, no. 11 (2014): 2150–2206, https://doi.org/10.1016/j.clinph.2014.05.021.

282 F. S. Bersani et al., "Deep Transcranial Magnetic Stimulation as a Treatment for Psychiatric Disorders: A Comprehensive Review," *European Psychiatry* 28, no. 1 (2013): 30–39, https://doi.org/10.1016/j.eurpsy.2012.02.006.

283 Lefaucheur et al., "Evidence-Based Guidelines on the Therapeutic Use of Repetitive Transcranial Magnetic Stimulation," 2150–2206.

284 S. Rossi et al., "Safety and Recommendations for TMS Use in Healthy Subjects and Patient Populations, with Updates on Training, Ethical and Regulatory Issues: Expert Guidelines," *Clinical Neurophysiology* 132, no. 1 (2021): 269–306, https://doi.org/10.1016/j.clinph.2020.10.003.

285 W. M. Gregory et al., "Frequency-Specific Microcurrent Improves Hand Function and Raynaud's Symptoms in Scleroderma: Results of Two Pilot Studies," *Rheumatology (Oxford)* 64, no. 10 (2025): 5504–8, https://doi.org/10.1093/rheumatology/keaf301.

286 J. C. Nielsen et al., "Radiofrequency Ablation as Initial Therapy in Paroxysmal Atrial Fibrillation," *The New England Journal of Medicine* 367 (2012): 1587–95, https://doi.org/10.1056/NEJMoa1113566.

287 Y. Ni et al., "A Review of the General Aspects of Radiofrequency Ablation," *Abdominal Imaging* 30, no. 4 (2005): 381–400, https://doi.org/10.1007/s00261-004-0253-9.

288 D.-K. Li et al., "Exposure to Magnetic Field Non-Ionizing Radiation and the Risk of Miscarriage: A Prospective Cohort Study," *Scientific Reports* 7 (2017): 17541, https://doi.org/10.1038/s41598-017-16623-8.

289 T. Sadeghi et al., "Preterm Birth Among Women Living Within 600 Meters of High Voltage Overhead Power Lines: A Case-Control Study," *Romanian Journal of Internal Medicine* 55, no. 3 (September 26, 2017): 145–50, https://doi.org/10.1515/rjim-2017-0017.

290 Sadeghi et al., "Preterm Birth Among Women Living Within 600 Meters of High Voltage Overhead Power Lines."

291 N. Auger et al., "Maternal Proximity to Extremely Low Frequency Electromagnetic Fields and Risk of Birth Defects," *European Journal of Epidemiology* 34 (2019): 689–97, https://doi.org/10.1007/s10654-019-00518-1.

292 D.-K. Li et al., "Maternal Exposure to Magnetic Fields During Pregnancy in Relation to the Risk of Asthma in Offspring," *Archives of Pediatric and Adolescent Medicine* 165, no. 10 (2011): 945–50, https://doi.org/10.1001/archpediatrics.2011.135.

293 Li et al., "Maternal Exposure to Magnetic Fields During Pregnancy."

294 D.-K. Li et al., "A Prospective Study of In-Utero Exposure to Magnetic Fields and the Risk of Childhood Obesity," *Scientific Reports* 2 (2012): 540, https://doi.org/10.1038/srep00540.

295 "Notice of Retraction and Replacement: Li et al., Association Between Maternal Exposure to Magnetic Field Nonionizing Radiation During Pregnancy and Risk of Attention-Deficit/Hyperactivity Disorder in Offspring in a Longitudinal Birth Cohort," *JAMA Network Open* 4, no. 2 (2021): e2033605, https://doi.org/10.1001/jamanetworkopen.2020.33605.

296 I. Al-Jarrah and M. Rababa, "Retraction Notice to 'Impacts of Smartphone Radiation on Pregnancy: A Systematic Review,'" *Heliyon* 9, no. 6 (2023): e16632, https://doi.org/10.1016/j.heliyon.2023.e16632.

297 M. E. Elkafrawy and D. M. Effat, "Effect of Cellular Phone Use on Fetal Monitoring and Umbilical Artery Doppler," *Annals of RSCB* 25, no. 6 (2021): 15826–35, https://doi.org/10.24941/ijcr.41956.07.2021.

298 T. S. Aldad et al., "Fetal Radiofrequency Radiation Exposure from 800–1900 MHz-Related Cellular Telephones Affects Neurodevelopment and Behavior in Mice," *Scientific Reports* 2 (2012): 312, https://doi.org/10.1038/srep00312.

299 L. Birks et al., "Maternal Cell Phone Use During Pregnancy and Child Behavioral Problems in Five Birth Cohorts," *Environment International* 104 (2017): 122–31, https://doi.org/10.1016/j.envint.2017.03.024.

300 M. Sudan et al., "Maternal Cell Phone Use During Pregnancy and Child Cognition at Age 5 Years in 3 Birth Cohorts," *Environment International* 120 (2018): 155–62, https://doi.org/10.1016/j.envint.2018.07.043.

301 C. V. Bellieni et al., "Electromagnetic Fields Produced by Incubators Influence Heart Rate Variability in Newborns," *Archives of Disease in Childhood-Fetal and Neonatal Edition* 93, no. 4 (2008): F298–F301, https://doi.org/10.1136/adc.2007.132738.

302 Bellieni et al., "Electromagnetic Fields Produced by Incubators Influence Heart Rate Variability."

303 S. Jullien, "Sudden Infant Death Syndrome Prevention," *BMC Pediatrics* 21, no. 2 (2021): 320, https://doi.org/10.1186/s12887-021-02536-z.

304 S. Bhattacharjee, "Protective Measures to Minimize the Electromagnetic Radiation," *Advance in Electronic and Electric Engineering* 4, no. 4 (2014): 375–80, https://ripublication.com/aeee_spl/aeeev4n4spl_08.pdf.

305 Sudan et al., "Maternal Cell Phone Use During Pregnancy and Child Cognition," 155–62.

306 H. A. Divan et al., "Cell Phone Use and Behavioural Problems in Young Children," *Journal of Epidemiology and Community Health* 66, no. 6 (2012): 524–29, https://doi.org/10.1136/jech.2010.115402.

307 A. Huss et al., "Environmental Radiofrequency Electromagnetic Fields Exposure at Home, Mobile and Cordless Phone Use, and Sleep Problems in 7-Year-Old Children," *PLoS One* 10, no. 10 (2015): e0139869, https://doi.org/10.1371/journal.pone.0139869.

308 M. Foerster et al., "A Prospective Cohort Study of Adolescents' Memory Performance and Individual Brain Dose of Microwave Radiation from Wireless Communication," *Environmental Health Perspectives* 126, no. 7 (2018): 077007, https://doi.org/10.1289/ehp2427.

309 J. E. Costello et al., "Is There an Epidemic of Child or Adolescent Depression?" *Journal of Child Psychology and Psychiatry* 47, no. 12 (2006): 1263–71, https://doi.org/10.1111/j.1469-7610.2006.01682.x.

310 R. Mojtabai et al., "National Trends in the Prevalence and Treatment of Depression in Adolescents and Young Adults," *Pediatrics* 138, no. 6 (2016): e20161878, https://doi.org/10.1542/peds.2016-1878.

311 "2020 National Survey on Drug Use and Health (NSDUH)," Substance Abuse and Mental Health Services Administration, https://www.samhsa.gov/data/data-we-collect/nsduh-national-survey-drug-use-and-health/national-releases/2020.

312 G. Lissak, "Adverse Physiological and Psychological Effects of Screen Time on Children and Adolescents: Literature Review and Case Study," *Environmental Research* 164 (2018): 149–57, https://doi.org/10.1016/j.envres.2018.01.015.

313 D. Maras et al., "Screen Time Is Associated with Depression and Anxiety in Canadian Youth," *Preventive Medicine* 73 (2015): 133–38, https://doi.org/10.1016/j.ypmed.2015.01.029.

314 C. Fernández et al., "Absorption of Wireless Radiation in the Child Versus Adult Brain and Eye from Cell Phone Conversation or Virtual Reality," *Environmental Research* 167 (2018): 694–99, https://doi.org/10.1016/j.envres.2018.05.013.

315 "Cell Phone Radiation & Children's Health: What Parents Need to Know," HealthyChildren.org, American Academy of Pediatrics, accessed October 11, 2023, https://www.healthychildren.org/English/safety-prevention/all-around/Pages/Cell-Phone-Radiation-Childrens-Health.aspx.

316 *Final Report on Commission to Study the Environmental and Health Effects of Evolving 5G Technology (RSA 12-K:12-13, HB-522, Ch. 260, Laws of 2019)* (State of New Hampshire, November 1, 2020), https://www.gencourt.state.nh.us/statstudcomm/committees/1474/reports/5G%20final%20report.pdf.

317 HB 548, "EDUCATION: Requires the State Department of Education to Develop and Distribute Health and Safety Guidelines Relative to Best Practices for the Use of Digital Devices in Public Schools," Louisiana State Legislature, May 31, 2022, https://legis.la.gov/legis/BillInfo.aspx?s=22RS&b=HB548&sbi=y.

318 Reuters in Amsterdam, "Mobile Phones and Other Devices to Be Banned from Dutch Classrooms," *The Guardian*, July 4, 2023, https://www.theguardian.com/world/2023/jul/04/mobile-phones-other-devices-to-be-banned-from-dutch-classrooms.

319 V. I. Mordachev, "Correlation Between the Potential Electromagnetic Pollution Level and the Danger of COVID-19: 4G/5G/6G Can Be Safe for People," *Doklady BGUIR* 18, no. 4 (2020): 96–112, https://doi.org/10.35596/1729-7648-2020-18-4-96-112.

320 A. Tsiang and M. Havas, "COVID-19 Attributed Cases and Deaths Are Statistically Higher in States and Counties with 5th Generation Millimeter Wave Wireless Telecommunications in the United States," *Medical Research Archives* 9, no. 4 (2021), https://doi.org/10.18103/mra.v9i4.2371.

321 N. Alster, *Captured Agency: How the Federal Communications Commission Is Dominated by the Industries It Presumably Regulates* (Edmond J. Safra

Center for Ethics, Harvard University), https://ethics.harvard.edu/files/center-for-ethics/files/capturedagency_alster.pdf.

322 "D.C. Circuit Decision—Environmental Health Trust v. FCC," August 16, 2021, https://www.fcc.gov/document/dc-circuit-decision-environmental-health-trust-v-fcc.

323 S. Cucurachi et al., "A Review of the Ecological Effects of Radiofrequency Electromagnetic Fields (RF-EMF)," *Environment International* 51 (2013): 116–40, https://doi.org/10.1016/j.envint.2012.10.009.

324 J. Klune et al., "Tracking Devices for Pets: Health Assessment for Exposure to Radiofrequency Electromagnetic Fields," *Animals* 11, no. 9 (2021): 2721, https://doi.org/10.3390/ani11092721.

325 M. Rao et al., "Experimental Evidence of Radio Frequency Radiation from *Staphylococcus aureus* Biofilms," *IEEE Journal of Electromagnetics, RF and Microwaves in Medicine and Biology* 6, no. 3 (2022): 420–28, https://doi.org/10.1109/JERM.2022.3168618.